AF595330

Fiber Optic Sensors in Chemical and Biological Applications

Fiber Optic Sensors in Chemical and Biological Applications

Editor

Gabriela Kuncová

MDPI • Basel • Beijing • Wuhan • Barcelona • Belgrade • Manchester • Tokyo • Cluj • Tianjin

Editor
Gabriela Kuncová
Faculty of Environment
University of Jan Evangelista
Purkyne (UJEP)
Czech Republic

Editorial Office
MDPI
St. Alban-Anlage 66
4052 Basel, Switzerland

This is a reprint of articles from the Special Issue published online in the open access journal *Sensors* (ISSN 1424-8220) (available at: https://www.mdpi.com/journal/sensors/special_issues/Fiber_Optic_Sensors_Chemical_Biological).

For citation purposes, cite each article independently as indicated on the article page online and as indicated below:

LastName, A.A.; LastName, B.B.; LastName, C.C. Article Title. *Journal Name* **Year**, *Volume Number*, Page Range.

ISBN 978-3-0365-4157-0 (Hbk)
ISBN 978-3-0365-4158-7 (PDF)

Contents

About the Editor

Gabriela Kuncova

Dr. Gabriela Kuncova is a lecturer at the University of Jan Evangelista Purkyne in Usti nad Labem, and an Emerita scientist at the Institute of Chemical Process Fundamentals (ICPF) Academy of Sciences in the Czech Republic, Prague. From 1979 to 1989, she participated in the research and development of drawing and coating optical fibers. From 1998 to 2017, she was the head of the group of Immobilized Biomaterials and Optical Sensors at ICPF. The use of optical fiber sensors in bioreactors, and for monitoring environmental pollution, immobilization of biological materials into organic–inorganic matrices, enzyme and whole-cell biosensors, and biodegradation with immobilized biocatalysts have been her main research interests. She is an author or co-author of 79 registered papers (740 citations without self-citations (Web of Science)), with an H index of 20.

Preface to "Fiber Optic Sensors in Chemical and Biological Applications"

Research on optical fiber sensors for chemical and biological applications began soon after the production of optical fibers in 1970, and boomed at the end of the second millennium. At the beginning, optical fibers served to explore light transmission from a measured area to a detector, and research was focused on the fixation of known chemical indicators and/or sensitive biological parts to the surface of an optical fiber. Currently, such types of sensors may be used in harsh conditions, particularly in reactors. The miniaturization and availability of advanced electronic devices are the reasons why extrinsic optical fiber chemical/bio-sensors have often been superseded by nanosensors and optical sensors with wireless transmitters, above all, in medical and environmental applications. Thus, recent research pertaining to optical fiber sensors in chemical and biological applications has mostly been oriented around intrinsic optical fiber sensors, whereby the waveguide structure of the optical fibers plays an active role in sensing. Such label-free chemical/bio-sensors are highly sensitive; however, their low selectivities limit their practical use to very specific analyses, for the detection of impurities.

In this Special Issue, the subjects of the papers truly reflect the contemporary stage of research interests in the field of chemical and biological applications of optical fiber sensors.

Gabriela Kuncová
Editor

MDPI

Article

Fabrication of Humidity-Resistant Optical Fiber Sensor for Ammonia Sensing Using Diazo Resin-Photocrosslinked Films with a Porphyrin-Polystyrene Binary Mixture

Soad Ahmed [1], Yeawon Park [1], Hirofumi Okuda [1], Shoichiro Ono [1], Sergiy Korposh [2] and Seung-Woo Lee [1,*]

[1] Graduate School of Environmental Engineering, The University of Kitakyushu, 1-1 Hibikino, Kitakyushu 808-0135, Japan; z8daa401@eng.kitakyu-u.ac.jp (S.A.); b0daa001@eng.kitakyu-u.ac.jp (Y.P.); okuda-hirofumi@toyoko-jp.com (H.O.); a9maa006@eng.kitakyu-u.ac.jp (S.O.)

[2] Department of Electrical and Electronic Engineering, University of Nottingham, Nottingham NG7 2RD, UK; s.korposh@nottingham.ac.uk

* Correspondence: leesw@kitakyu-u.ac.jp; Tel.: +81-93-695-3293

Abstract: Ammonia gas sensors were fabricated via layer-by-layer (LbL) deposition of diazo resin (DAR) and a binary mixture of tetrakis(4-sulfophenyl)porphine (TSPP) and poly(styrene sulfonate) (PSS) onto the core of a multimode U-bent optical fiber. The penetration of light transferred into the evanescent field was enhanced by stripping the polymer cladding and coating the fiber core. The electrostatic interaction between the diazonium ion in DAR and the sulfonate residues in TSPP and PSS was converted into covalent bonds using UV irradiation. The photoreaction between the layers was confirmed by UV-vis and Fourier transform infrared spectroscopy. The sensitivity of the optical fiber sensors to ammonia was linear when exposed to ammonia gases generated from aqueous ammonia solutions at a concentration of approximately 17 parts per million (ppm). This linearity extended up to 50 ppm when the exposure time (30 s) was shortened. The response and recovery times were reduced to 30 s with a 5-cycle DAR/TSPP+PSS (as a mixture of 1 mM TSPP and 0.025 wt% PSS in water) film sensor. The limit of detection (LOD) of the optimized sensor was estimated to be 0.31 ppm for ammonia in solution, corresponding to approximately 0.03 ppm of ammonia gas. It is hypothesized that the presence of the hydrophobic moiety of PSS in the matrix suppressed the effects of humidity on the sensor response. The sensor response was stable and reproducible over seven days. The PSS-containing U-bent fiber sensor also showed superior sensitivity to ammonia when examined alongside amine and non-amine analytes.

Keywords: ammonia detection; layer-by-layer; U-bent optical fiber; porphyrin; poly(styrene sulfonate); diazo resin; photocrosslinking

Citation: Ahmed, S.; Park, Y.; Okuda, H.; Ono, S.; Korposh, S.; Lee, S.-W. Fabrication of Humidity-Resistant Optical Fiber Sensor for Ammonia Sensing Using Diazo Resin-Photocrosslinked Films with a Porphyrin-Polystyrene Binary Mixture. *Sensors* **2021**, *21*, 6176. https://doi.org/10.3390/s21186176

Academic Editor: Gabriela Kuncová

Received: 21 May 2021
Accepted: 19 June 2021
Published: 15 September 2021

1. Introduction

The use of optical fibers for chemical and biological sensing has become popular for both scientific and commercial applications [1,2]. Because of their unique properties, such as small size, immunity to electromagnetic interference, and biocompatibility, they facilitate remote and real-time measurements with high sensitivity and selectivity. Most importantly, an optical fiber sensor (OFS) can be deployed in a harsh corrosive environment owing to the chemically inert silica substrate of the fiber [3].

The different OFS geometries reported in the literature for refractive index (RI) and absorption sensings include straight decladded [4], partially polished [5], laterally polished [6], D-shaped [7], and U-bent [8] fibers. Among the various reported designs, the U-bent design has several advantages, such as (1) a high evanescent wave absorbance sensitivity due to conversion of the lower-order modes into higher-order modes, (2) less fragility compared to other geometries, (3) ease of probe fabrication and repeatability, and (4) ease of development into a point sensor. Combining the advantages of the U-bent geometry and the stripped fiber core functionalized with nanomaterials enables the development of a high-sensitivity sensor capable of binding target chemical species.

Ammonia (CAS No. 7664-41-7) is a widely used chemical in the chemical industry. It is an important gas for assessing indoor air quality, particularly in the industrial sector, as its presence in excess of its exposure limit results in health issues [9]. Its permissible exposure limit is regulated to be 50 ppm (35 mg m^3), according to the Occupational Safety and Health Administration (OSHA) guidelines [10]. On the other hand, ammonia has recently attracted a great deal of attention as a hydrogen carrier and an alternative energy resource to replace hydrogen [11]. From this perspective, the development of technologies related to ammonia gas detection, separation, condensation, etc., becomes an important issue. Moreover, the use of ammonia as a biomarker in human breath [12,13] for detecting pathological disorders such as renal insufficiency [14], hepatic dysfunction [15], *Helicobacter pylori* infection [16], and halitosis [17], has recently gained significant interest from researchers [18].

From this perspective, OFSs provide an excellent base for developing low-cost, small, sensitive, and reliable ammonia sensors, making them viable for on-field applications. In the past decades, sensing agents made of various organic polymers, dyes, and pH indicators using OFSs have been investigated for detecting ammonia. In these cases, the ammonia reacts with the water molecules to produce hydroxide ions (OH^-), which deprotonate the sensing agent, resulting in a change in the absorption spectrum of the film [19].

In our previous work, we demonstrated evanescent-wave-based OFSs for ammonia gas sensing, which were modified with tetrakis(4-sulfophenyl)porphine (TSPP) and several cationic polymers via electrostatic interactions between the oppositely charged polyelectrolytes [20–22]. The exposure of the TSPP nanonassembled film to ammonia induced unique optical changes in the transmission spectrum of the optical fiber, displaying the characteristic absorption bands (two Soret bands and Q bands) of the assembled TSPP compound [20]. A high extinction coefficient (>2.0 $\times$ 10^5 cm^{-1} M^{-1}) makes porphyrin particularly viable for the creation of optical sensors. However, several issues challenge its implementation when used with optical fibers, such as leaching out of the TSPP from the nanoassembled films [21], undesirable sensitivity to humidity owing to the hydrophilicity of the porphyrin, swelling of the immobilized polymer, long recovery time at high ammonia concentrations (>1 ppm), and a longer requisite exposure time (>30 s) [22].

A solution to such problems is creating covalent bonds or effective interactions between the dyes and the embedding matrix. It has been reported that polyelectrolyte complexes that are formed using diazo resin (DAR) as a cationic polyelectrolyte can be converted into covalent bonds by irradiation with ultraviolet (UV) light [23]. Crosslinked DAR films are stable even at high pH due to their high resistance to solvent etching, resulting in denser and more rigid films [24].

The present study proposes a novel strategy for fabricating a U-bent OFS coated with covalently crosslinked porphyrin layers. As mentioned above, the U-bent fiber geometry is advantageous in improved evanescent wave interactions due to the ergonomic design of the fiber compared to other fiber geometries. DAR as a polycation polymer and a binary mixture of TSPP and poly(styrene sulfonate) (PSS) as polyanions were alternately deposited as a film by the layer-by-layer (LbL) method on the U-bent fiber core. It is presumed that the presence of PSS in the matrix imparts greater stability to the low-molecular-weight dyes or pigments in the film and reduces the interference due to water on the sensor response. On the other hand, it may affect the mobility or aggregate structures of the TSPP, and thus, the amount of PSS employed in the film was optimized. Photopolymerization via UV irradiation resulted in a greater stability of the TSPP inside the film, which was co-assembled with PSS. Herein, the responses of the sensor to different concentrations of ammonia were measured under nearly saturated humidity conditions, and the critical sensor parameters, such as selectivity, sensitivity, and stability, were investigated.

2. Materials and Methods

2.1. Materials

TSPP (Mw: 934.99 g mol^{-1}) was purchased from Tokyo Chemical Industry Co., Ltd. (Tokyo, Japan). PSS (Mw: 70.000 g mol^{-1}) and acetone were purchased from Sigma-Aldrich (St. Louis, MO, USA). NH_4OH (28 wt% in H_2O), which was used to produce the ammonia analyte gas, and pyridine (Py), ethanol, toluene, methanol, and 1-hexanol, which were used as additional analyte gases, were purchased from Wako Pure Chemical Industries (Osaka, Japan). Trimethylamine (TMA, 30% in H_2O) and 1-propanol were purchased from Kanto Chemical Co., Ltd. (Osaka, Japan). Triethylamine (TEA) was purchased from Kishida Chemical Co., Ltd. (Osaka, Japan). All chemicals were of analytical grade and were used without further purification. DAR (Mw: ca. 2500 g mol^{-1}) was synthesized according to a previously reported method [25,26]. Deionized (DI) pure water (18.2 MΩ·cm) was obtained by reverse osmosis, followed by ion exchange and filtration (Aquapuri 541; Young In Scientific, Seoul, Korea).

2.2. Preparation of Optical Fiber Sensor

To fabricate the OFS, a short section of the plastic cladding of a multimode optical fiber (HWF 200/230/500T silica core/plastic cladding with a 200 µm core diameter, CeramOptec: Bonn, Germany) was burned and bent to a U-shape using a burner. The total length of the fabricated U-bent OFS was approximately 40 cm, and the exposed length of the U-shaped tip probe (without plastic cladding) was approximately 1 cm, as shown in Figure 1a. Then, the exposed section of the silica core was rinsed in ethanol and DI water several times, and treated with 1 wt% ethanolic KOH (ethanol/water = 3:2, v/v) for 20 min. Thus, the negatively charged evanescent region was formed.

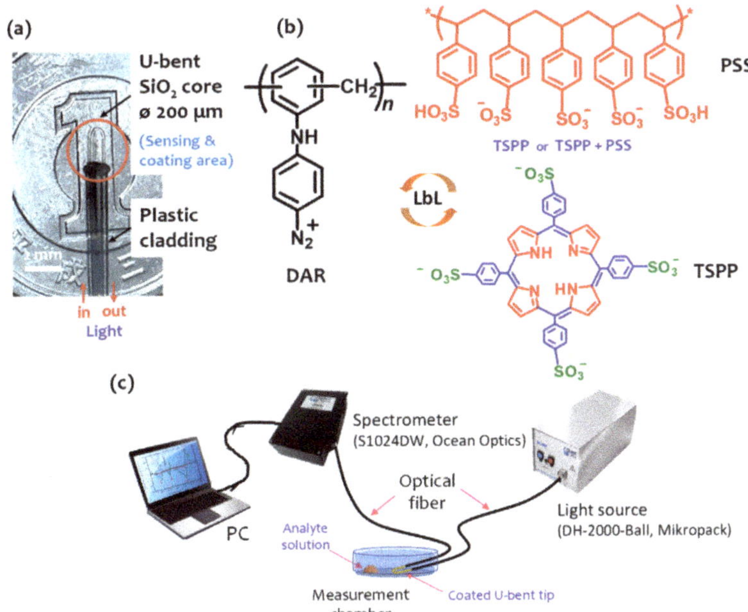

Figure 1. (**a**) Photograph of a U-bent optical fiber. (**b**) Chemical structures of DAR, TSPP, and PSS and a schematic of the LbL assembly of DAR and TSPP, or DAR and a TSPP+PSS binary mixture on a multimode U-bent optical fiber core followed by exposure to UV-radiation at 365 nm. (c) Schematic of the experimental setup for gas detection.

Next, an electrostatic film was deposited by alternately immersing the negatively charged activated core of the optical fiber into an aqueous DAR solution (1 wt%, pH 2.2) and a mixture of TSPP (1 mM in H_2O, pH 3) and PSS (0, 0, 0.025, and 0.1 wt% in H_2O) at room temperature for 15 min each, producing a DAR/TSPP+PSS bilayer (where one cycle results in a DAR/TSPP or DAR/TSPP+PSS bilayer). This was achieved by adding the coating solution (volume 70 μL) to a deposition cell with the intermediate processes of water washing and drying by flushing with nitrogen gas being. Multilayer assemblies can be formed by cyclic repetition of these two steps. Figure 1b shows the LbL assembly of the photocrosslinkable DAR and TSPP or a binary mixture of TSPP and PSS. An LbL film without PSS was prepared to evaluate the effect of PSS on the sensor response, and the outermost layer of the alternate film was DAR in all cases. The films were deposited in the dark to avoid the decomposition of the DAR. Finally, the U-bent optical fiber modified with a 5-cycle DAR/TSPP or DAR/TSPP+PSS thin film was cured by exposing it to UV light (λ_{max} = 365 nm) at a distance of 10 cm for 15 min. Consequently, the ionic bonds between the diazonium cationic ions in DAR and the sulfonate anionic residues in TSPP and PSS can be converted into crosslinked covalent bonds by UV light irradiation.

2.3. Experimental Setup and Sensing Performance

The stripped section of the optical fiber that was coated with a functional film was fixed in a specially designed cylindrical acrylic gas measurement chamber (3.6 cm in diameter, 1.0 cm in height; internal volume: 10.2 cm^3). As seen in Figure 1c, one end of the optical fiber was connected to a deuterium halogen light source (DH-2000-Ball, Mikropack), while the other end was connected to a spectrometer (S1024DW, Ocean Optics) to monitor the assembly process and gas sensing. Spectra Suite® Spectrometer Operating Software (Ocean Optics) was used for the analysis. The absorbance (A) was determined by taking the logarithm of the ratio of the transmission spectrum of the coated fiber, T(λ), to the transmission spectrum measured prior to the film deposition, $T_0(\lambda)$, as shown in Equation (1):

$$A(\lambda) = -\log\left[\frac{T(\lambda)}{T_0(\lambda)}\right]. \quad (1)$$

The response of the U-bent OFS was measured by exposing it to ammonia and other analyte gases at different concentrations, which were generated by placing a fixed volume (100 μL) of their aqueous analyte solutions of different concentrations inside the measurement chamber near the U-bent tip. Each analyte solution was kept for approximately 3 min to attain equilibrium between the released gas and the solution (also kept until the saturation of the signals measured at the wavelength of 706 nm). The individual amine gas concentrations of ammonia, TMA, TEA, and Py were measured using gas tubes (No. 3L for ammonia, No. 180 for TMA, No. 180L for TEA, and No. 182 for Py; GasTech, Inc., Ayase, Japan). The actual amine gas concentrations were estimated using calibration curves obtained from the corresponding amine concentrations in the solutions, as shown in Figure S1.

The baseline spectrum of each experiment was measured by placing DI water (100 μL) at the bottom of the measurement chamber near the tip of the U-bent optical fiber until the signals measured at the wavelength of 706 nm stabilized. The relative humidity level inside the measurement chamber during all measurements reached approximately 80% at room temperature (approximately 24 °C) (Figure S2). The sensor response (SR) was calculated using the following equation Equation (2):

$$SR\ (\%) = \frac{I_0 - I}{I_0} \times 100, \quad (2)$$

where I_0 and I are the light intensities of the OFS in the absence and presence of an analyte gas, respectively, measured at the same wavelength.

2.4. Film Characterization

Scanning electron microscopy (SEM) measurements were performed with a Hitachi S-5200 at an acceleration voltage of 10–15 kV. A 2-nm thick platinum layer was deposited on all samples using a Hitachi E-1030 ion sputter at 15 mA and 10 Pa prior to measurements, preventing the charge up of the sample surface. For Fourier transform infrared spectroscopy (FT-IR) analysis, the films were prepared on a gold-coated silicon wafer plate and measured using a Spectrum 100 FT-IR spectrometer (Perkin Elmer Japan Co., Ltd., Yokohama, Japan). The contact angles of the samples were measured using a DropMaster 100 (DropMaster Co., Ltd., Niiza, Japan).

3. Results and Discussion

3.1. Strategy for Sensitive and Reproducible Sensor Fabrication

TSPP, which has two distinct and unique aggregates (J- and H-) along with monomers in nanoassembled films, has been extensively investigated in previous studies, including our trials [21,27,28]. The concentrations of these structures usually change with the degree of TSPP protonation or deprotonation [20,29]. These optical features can be used for the detection of some analytes that are proton-donating or proton-accepting in aqueous or gaseous media [22,27]. However, the TSPP aggregations inside the film are strongly affected by humidity because of the hydrophilicity of TSPP. As a result, the water molecules trapped in the film matrix retain the analyte for an extended period. This drawback of the nanoassembled TSPP films for chemical sensing may be caused by the fundamental features of the layered film, which are based on the electrostatic interaction between the alternate layers.

DAR is a candidate molecule for solving this problem because it is capable of changing the electrostatic bond between the layers to a covalent bond. The arene diazonium salt (DAR-N2+) ion-complexed with an anionic species can be easily decomposed via UV irradiation or heat treatment by releasing the N2 gas. Consequently, the negatively charged sulfonate groups of TSPP can be covalently crosslinked through nucleophilic addition to the carbocation formed in the terminal benzene ring of the DAR. Sun et al. reported that negatively charged TSPP sulfonate ions complexed with DAR can be converted into covalent bonds after photoreaction induced by UV irradiation [30]. In this study, a novel approach was demonstrated to improve the hydrophobicity of the sensor films by using a binary mixture of TSPP and PSS. PSS is an anionic polymer having sulfonate groups and makes strong ion-complexes with diazonium functional groups. As illustrated in Figure 1b, the assembled layers can be covalently crosslinked after UV irradiation and become much more hydrophobic due to the neutrally transformed phenyl rings in the PSS. However, the content of PSS employed in the sensor films may competitively affect the sensing ability of TSPP.

3.2. Optical Features of Covalently Attached DAR/TSPP+PSS Multilayers

The layered film assembly with DAR and TSPP or DAR and a binary mixture of TSPP and PSS was confirmed after each deposition cycle by monitoring optical changes in the transmission spectra of the optical fiber. Optical fiber absorption spectra were obtained by adapting Equation (1) to the transmission spectra recorded during the film deposition (Figure S3). Figure 2a,b show the absorption spectra of a 5-cycle DAR/TSPP+PSS film with an outermost layer of TSPP+PSS and DAR, respectively, where the PSS concentration in the coating solution was adjusted to 0.025 wt%. This PSS concentration was optimal for improving sensitivity to ammonia and hydrophobicity of the sensor film. Figure 2c shows the absorbance changes at 435, 490, and 706 nm during the film deposition when the outermost layer was deposited with a TSPP+PSS binary mixture. The absorbance at each wavelength increased linearly with the number of layers, indicating the uniform film deposition for each component. The absorption spectra for the individual layers with a TSPP+PSS outermost layer were characterized by a double peak in the Soret band at 435 and 490 nm and a pronounced Q band at approximately 706 nm. In addition to these characteristic peaks, the absorbances at 380 and 309 nm correspond to the π–π^* transition of the diazonium group [30] and the benzene rings of DAR and PSS, respectively. Notably, no significant changes in the position or shape of the TSPP absorption peaks were observed when the outermost layer was deposited with DAR, which indicates that the TSPP aggregate structures do not change when overlaid with DAR layers. Moreover, mixing TSPP with a small amount of PSS did not significantly affect the absorption spectra of the film.

The decomposition of the diazonium groups in the 5-bilayer DAR/TSPP+PSS (0.025 wt%) film completed within 15 min. Figure 2d shows a decrease in absorbance at 380 nm, indicating that the diazonium groups decomposed under UV irradiation. Simultaneously, the absorbance of the Soret band of TSPP at approximately 435 nm also slightly decreased without any change in the peak position and shape. This small decrease at 435 nm may be due to masking of the diazonium group spectra. Therefore, we concluded that the absorbance at 435 nm is mainly composed of the Soret band of the TSPP but partly overlaps with the edge of the diazonium group band before UV irradiation. This result also suggests that no significant changes occurred in the film conformation during the photocrosslinking reaction, as reported in a previous study [24]. On the other hand, the Q band of TSPP was blue-shifted (ca. 3 nm) with a slight increase in the absorbance at 706 nm, indicating that UV irradiation may induce structural changes in TSPP J-aggregates.

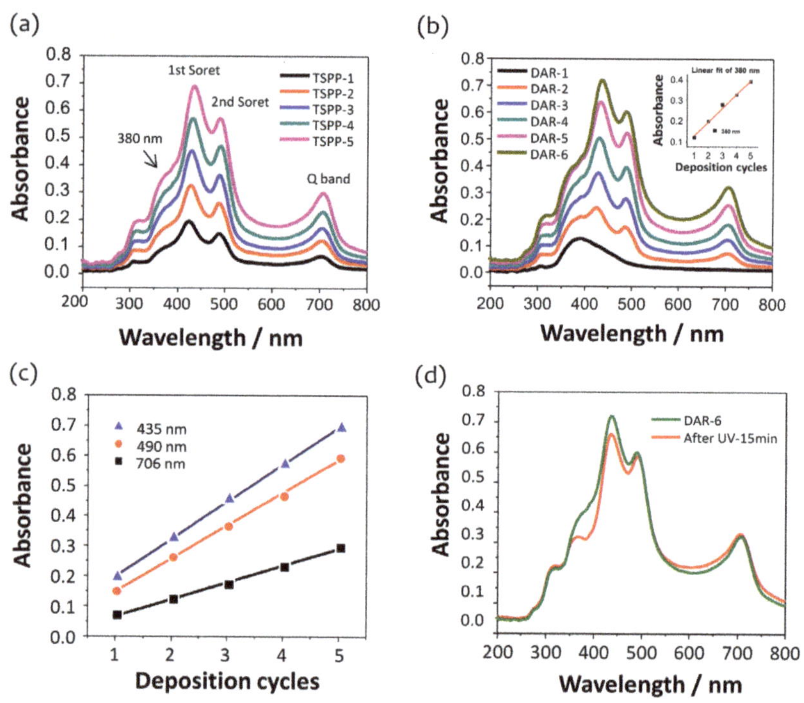

Figure 2. Evolution of the UV-vis absorption spectra of the DAR/TSPP+PSS (0.025 wt%) alternate layers deposited onto the 1-cm-long stripped core of a U-bent optical fiber when the outermost layer was deposited with (**a**) TSPP+PSS and (**b**) DAR, respectively. (**c**) Absorbance changes measured at (▲) 435, (•) 490, and (■) 706 nm. (**d**) Comparison of the UV-vis absorption spectra of the 5-cycle DAR/TSPP+PSS (0.025 wt%) film with an outermost layer of DAR before and after exposure to UV irradiation for 15 min.

3.3. Sensor Responses to Ammonia

According to the equilibrium equation (Equation (3)), the ammonia gas that evaporated from the aqueous ammonia solutions saturated inside the measurement chamber.

$$NH_4OH(aq) \leftrightarrows NH_3(g) + H_2O \tag{3}$$

Figure 3a shows the intensity changes (defined as ΔI) in the transmission spectra due to the presence of ammonia in the 5-cycle DAR/TSPP+PSS (0.025 wt%) film, which is obtained by subtracting the transmission spectrum measured at a given ammonia concentration from that measured using water. As the ammonia concentration increased from 0 ppm to 100 ppm (sol), the intensity changes in the transmission spectra increased at 485 and 706 nm and decreased at 411 and 518 nm, respectively, as shown in Figure 3b. The most perceptible change in the intensity of the transmission spectra was observed at 706 nm. This result suggests the disappearance of the electrostatic interactions between the TSPP molecules due to the deprotonation process induced by the adsorption of ammonia molecules [22,31], indicating the distortion of the TSPP J-aggregates in the film.

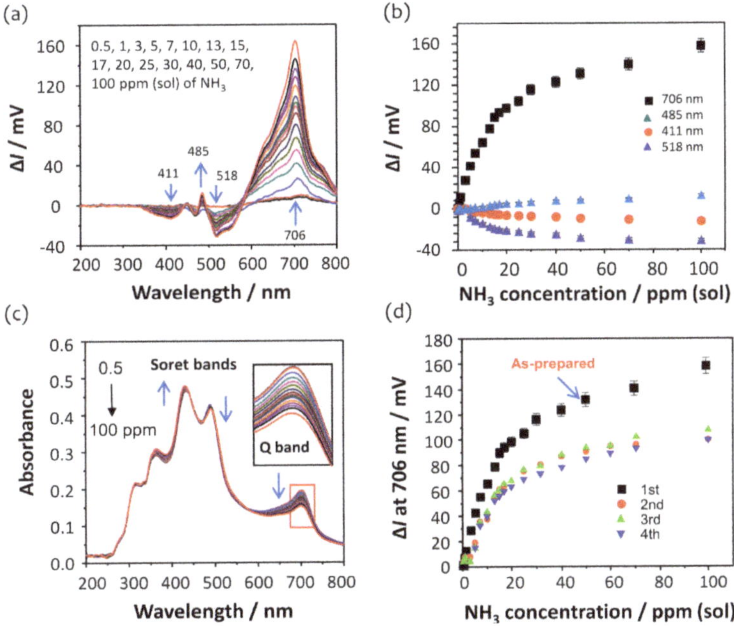

Figure 3. (**a**) Evolution of the transmission spectra induced by the exposure of the U-bent OFS modified with a 5-cycle DAR/TSPP+PSS (0.025 wt%) film to ammonia concentrations from 0 to 100 ppm (sol). (**b**) The intensity changes at four different wavelengths versus ammonia concentration, obtained from the three independent experiments conducted using the individual as-prepared sensors fabricated under the same conditions. (**c**) The corresponding UV-vis absorption spectral changes due to the exposure of the film to ammonia. The inset shows the decrease in absorbance of the Q band centered at 706 nm induced by the presence of ammonia. (**d**) The intensity changes at 706 nm, obtained from the four consecutive ammonia measurements.

As evident from Figure 3c, the Q band centered at 706 nm shows perceptible absorbance changes in the corresponding absorption spectra obtained using Equation (1), which are similar to the results observed in the transmission spectra shown in Figure 3a. The absorption peaks at 490 and 706 nm, which are attributed to the TSPP J-aggregates in the film, were reduced, while the absorbance of the peak at 435 nm increased.

Figure 3d shows the intensity changes at 706 nm, obtained from the four consecutive ammonia measurements in the concentration range from 0 to 100 ppm (sol). The intensity changes after the first measurement of the as-prepared film exhibited good reproducibility, with a standard deviation of ±4.5 mV at 100 ppm concentration of ammonia, although the average intensity change of ammonia measured thrice at 70 ppm (sol) was approximately 70% of the initial measurement. This decrease in intensity change after the first measurement of the as-prepared film may be due to the optimization of the film structure after continuous use. The higher amount of TSPP J-aggregates is present in the as-prepared film; however, some of the TSPP molecules, particularly those covalently bound to the DAR, are hard to return to the original J-aggregation state after the first use. The film matrix appears not to recover the initial state completely. Consequently, the TSPP J-aggregation is subtly realigned after the initial exposure to ammonia and optimized for ammonia sensing.

3.4. Optimization of the Content of PSS Employed in the Sensor Film

The reasons for employing PSS with TSPP in this study are as follows: first, a PSS polymer has multiple sulfonate functional groups and can be crosslinked with DAR upon UV irradiation, which helps form a covalently bonded stable film [23,24]; second, PSS may enhance the hydrophobicity of the film when added by mixing it with TSPP; the enhanced hydrophobicity of the film reduce the desorption of TSPP from it as well as its affinity to water molecules. These advantages of PSS improve the stability and sensitivity of the crosslinked film.

To investigate the effect of PSS in the sensor film on ammonia gas sensing under highly humid conditions, three types of covalently crosslinked DAR/TSPP+PSS films with different PSS contents of 0 wt%, 0.025 wt%, and 0.1 wt% were prepared. Figure 4a shows an example of the dynamic SRs measured at 706 nm for the U-bent OFS coated with a DAR/TSPP+PSS (0.025%) film when exposed to different ammonia concentrations from 0 to 100 ppm (sol). The SR of the sensor film was recorded for 3 min after placing 100 μL of aqueous ammonia of a given concentration in the chamber; the ammonia solution was then removed by suction and the chamber was flushed with dry air at a flow rate of 1 L min^{-1} for 1 to 2 min. The SR, calculated using Equation (2), was exceedingly quick (<30 s) for each ammonia solution and recovered almost entirely even after exposure to a high concentration of ammonia at 100 ppm (sol), exhibiting a SR of approximately 10%.

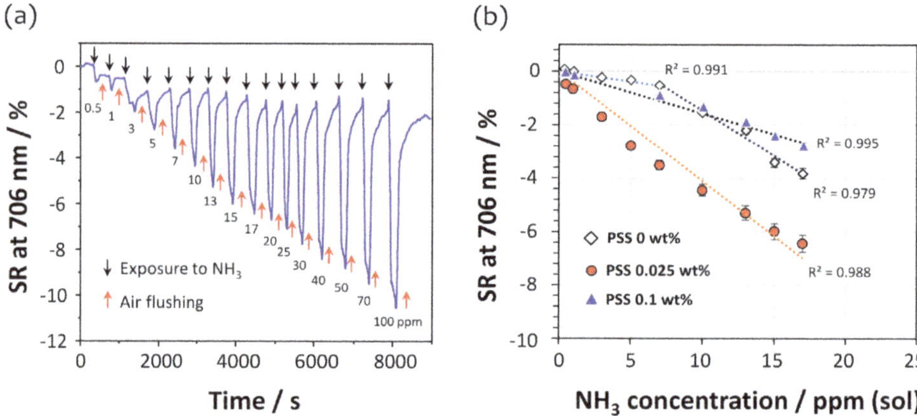

Figure 4. (**a**) Dynamic sensor responses (SRs) of the U-bent OFS coated with a 5-cycle DAR/TSPP+PSS (0.025 wt%) film at 706 nm when exposed to different ammonia concentrations from 0 to 100 ppm (sol). (**b**) SRs of the sensor films with different PSS contents of 0 wt%, 0.025 wt%, and 0.1 wt%, displaying a linear behavior within 17 ppm (sol) in each calibration curve.

Figure 4b compares the SRs of the three DAR/TSPP+PSS sensor films, which were measured at 706 nm in all cases. Interestingly, the SR was influenced by the PSS content and increased in the following order: PSS 0.1 wt%, PSS 0 wt%, and PSS 0.025 wt%. In particular, the sensor film without PSS exhibited a slightly different adsorption behavior, having two linear ranges (0–7 ppm and 7–17 ppm), compared to the other two sensor films containing PSS, indicating a slow reaction at low ammonia concentrations. It may be that the diffusion of ammonia into the film is disturbed at low concentrations.

3.5. Sensitivity and Selectivity of the PSS-Containing Sensor Films

As seen in Figure 4b, the linear behavior of the calibration curves for ammonia were observed within 20 ppm (sol). Interestingly, the film prepared with 0.025 wt% PSS exhibited a high SR compared to the other two films. Except for the sensor film without PSS, the SR of both PSS-containing sensor films to ammonia was almost linear with respect to the ammonia concentration in the range of 0–17 ppm (sol). It is presumed that embedding an appropriate amount of PSS inside the film increases the film's hydrophobicity and stability after photocrosslinking, suggesting that reduced water condensation inside the film allowed the permeation of analyte gases under highly humid conditions. As a result, the TSPP was able to repeatedly detect the presence of ammonia. Conversely, the rigidity of the film upon PSS reinforcement may hinder ammonia gas diffusion into the film matrix. Consequently, a decrease in sensitivity was observed when doping the TSPP layer with a higher PSS content (>0.1 wt%). Thus, the optimum PSS concentration was found to be 0.025 wt%. Details of the response and recovery times of the fiber sensors are listed in Table 1.

Table 1. Sensing parameters of the DAR/TSPP+PSS films with a different PSS content.

Film Name	Sensitivity (Slope) [a], % ppm^{-1} (R^2)	Response Time [b], s	Recovery Time, s	Linear Range [b], ppm (sol)	LOD (sol) [c], ppm (sol)
DAR/TSPP+PSS (0 wt%)	0.08 (0.991) 0.34 (0.979)	100	180	0–7 7–17	1.25 2.81
DAR/TSPP+PSS (0.025 wt%)	0.41 (0.988)	30	30	0–17	0.23
DAR/TSPP+PSS (0.1 wt%)	0.16 (0.997)	45	40	0–20	0.61

[a] Data measured at 120 s. [b] Response time determined as the interval (τ_{90}) needed for the signal to achieve 90% of its saturation measured at 50 ppm ammonia (sol). [c] Limit of detection (LOD) was estimated to be 0.096% SR as a 3σ level, where σ is 0.032% SR as a possible noise value.

The noise levels of the SRs were calculated from the baselines in a steady state before exposure to ammonia and were used to determine the limit of detection (LOD) of the U-bent fiber sensors for ammonia (data not shown). Three or more stable baselines were selected from the dynamic responses to ammonia in Figure 4a, and a value of 0.032 ± 0.002% SR was estimated as to be the noise value. Therefore, 0.096% SR at a 3 s level was used as the LOD. As described in Figure 4b and Table 1, the best ammonia sensing performance was observed in the DAR/TSPP+PSS (0.025 wt%) film, exhibiting an LOD of 0.23 ppm (sol) in addition to its fast response and recovery times (each 30 s). This LOD value is about three times smaller than that of the sensor film with 0.1 wt% PSS. Interestingly, the linear trend of the sensor response–concentration curve for the sensor with 0.025 wt% PSS was extended up to 50 ppm (sol) of ammonia when the exposure time (30 s) was shortened (Figure S4).

On the other hand, the curve for the sensor film without PSS displayed a shorter linear trend only in the concentration range of 0–7 ppm with an LOD of 1.25 ppm (sol) of ammonia. At a higher concentration range of 7–17 ppm, the film's response was enhanced, exhibiting a sensitivity of 0.34 ± 0.03% ppm^{-1}. This nonlinear sensitivity of the sensor film without PSS is probably because of a different diffusion mechanism of the ammonia into the film compared to that of both PSS-containing sensor films. Water molecules condensed on the surface of the sensor films disturb the diffusion of ammonia into the film owing to the high solubility of ammonia in water. Therefore, the hydrophobicity of PSS embedded in the film contributes to the efficient diffusion of ammonia gas at a high relative humidity.

An important sensor parameter is selectivity, that is, the capability to detect a particular substance from other coexisting substances. Therefore, the U-bent OFS coated with a 5-cycle DAR/TSPP+PSS (0.025 wt%) film was exposed to several amine and non-amine vapors along with ammonia. Figure 5a shows the calibration curves based on the intensity change, ΔI as sensor response, measured at 706 nm upon exposure to the vapors of ammonia, TMA, TEA, and Py, which were present individually or as a mixture of equal concentrations in the range of 0–17 ppm (sol). The DAR/TSPP+PSS (0.025 wt%) film exhibited the highest response to ammonia and then to TMA starting at 10 ppm (sol). In contrast, no sensitivity to TEA and Py was observed throughout the measured range (ΔI at 706 nm < 2 mV). The fiber sensor is most sensitive to ammonia among the four amine analytes; however, the sensitivity to ammonia decreased slightly in a mixture of the four amine analytes. This decrease means that the high sensitivity of the sensor to ammonia cannot be explained only by the basicity of the analytes alone (conjugate acid dissociation constant at logarithmic scale, p*K*a, at 25 °C: TEA, 10.75 > TMA, 9.87 > ammonia, 9.36 > Py, 5.25) [32]. An additional decisive parameter may be the vapor pressure of each analyte, which is shown in decreasing order: 6650 mmHg (21 °C) >> 430 mmHg (21 °C) > 51.75 mmHg (20 °C) ≅ 23.8 mmHg (25 °C for ammonia, TMA, TEA, and Py, respectively [32]. The SR to ammonia was four-fold larger than that to TMA at 15 ppm (sol) (ca. 60 and 15 mV for ammonia and TMA, respectively). In addition, the calibration range for TMA was narrow, ranging from 7 to 20 ppm (sol), compared to the concentration range of 0 to 17 ppm (sol) for ammonia.

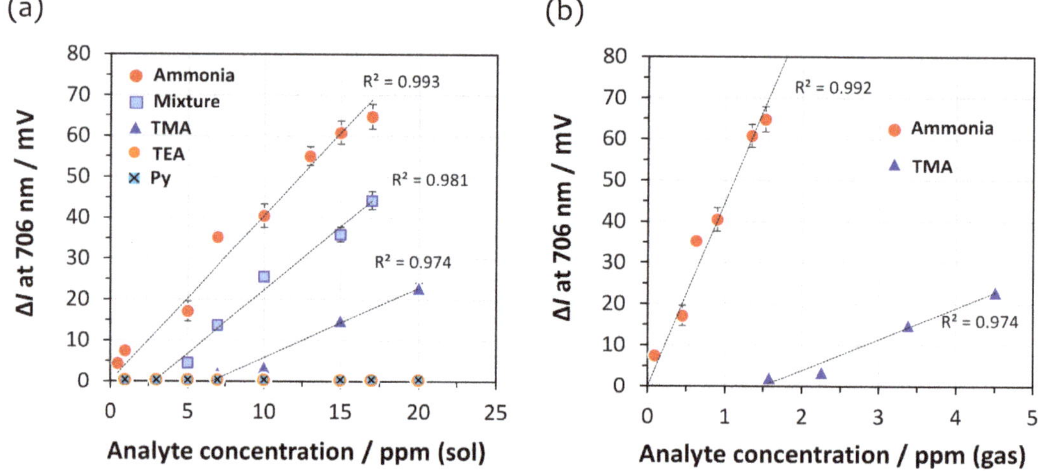

Figure 5. Calibration curves for the intensity change (ΔI as sensor response, $n = 3$) at 706 nm of the 5-cycle DAR/TSPP+PSS (0.025 wt%) film sensor (**a**) at different solution concentrations of ammonia, TMA, TEA, Py, and a mixture of all four analytes (equal concentrations in the range of 0–17 ppm for each), and (**b**) at different gas concentrations of ammonia and TMA.

Figure 5b shows the calibration curves based on the intensity change at 706 nm of the 5-cycle DAR/TSPP+PSS (0.025 wt%) film for different gas concentrations of ammonia and TMA. The actual gas concentrations of the four amine analytes (ammonia, TMA, TEA, and Py) vaporized from their corresponding aqueous solutions were measured using gas tubes (Figure S1). The linear correlation coefficients between the liquid and gas phase concentrations were estimated to be 0.090, 0.225, 0.059, and 0.188 for ammonia, TMA, TEA, and Py, respectively. It is presumed that the actual gas concentrations of TMA and Py are much higher than those of ammonia in the given concentration range. However, TMA was only detectable at gas concentrations over 1.5 ppm. Regardless of the relatively high gas concentrations of TMA and Py, the U-bent OFS surprisingly revealed the highest sensitivity to ammonia. Therefore, the decreased sensitivity in the case of the mixture of all four amines may be attributed to the interference from the less sensitive TMA and Py with relatively high gas concentrations.

Details of the sensitivity and LOD of the 5-cycle DAR/TSPP+PSS (0.025 wt%) film obtained when the solution and gas concentrations for ammonia and TMA were used are summarized in Table 2. When the actual gas concentrations for ammonia and TMA were used for sensitivity calculation, the LOD for ammonia was estimated to be about 0.03 ppm from the calibration curve in Figure 5b for the DAR/TSPP+PSS (0.025 wt%) film, whereas the LOD for TMA was 1.65 ppm, which is about 60 times the LOD for ammonia. From the calibration curves in Figure 5b, it is supposed that about 0.1 ppm of ammonia may be detectable even when 2 ppm of TMA coexists in gas samples. In practice, the LOD for the amine mixture is approximately ten times larger than that for ammonia, indirectly representing that the coexisting amines interfere slightly with the ammonia sensor response.

Table 2. Sensitivity and LOD of the DAR/TSPP+PSS (0.025 wt%) film obtained when the solution and gas concentrations for ammonia and TMA were used.

Analyte	Sensitivity (sol) [a], mV ppm^{-1} (R^2)	LOD (sol) [d], ppm	Sensitivity (gas), mV ppm^{-1}(R^2)	LOD (gas) [d], ppm
Ammonia	4.05 (0.993) [b]	0.31	44.7 (0.992)	2.85×10^{-2}
TMA	1.73 (0.974) [c]	7.21	7.56 (0.974)	1.65
Mixture	3.11 (0.974) [c]	3.23	NA	NA

[a] Data measured at 120 s. Applied concentration ranges: [b] 0–17 ppm and [c] 7–20 ppm. [d] LOD was estimated to be 1.28 mV as a 3σ level, where σ is 0.43 mV as a possible noise value (Figure S5). NA: not applicable.

On the other hand, this LOD (2.85×10^{-2} ppm) for ammonia is surprisingly 100 times smaller than that of the same 5-cycle PDDA/TSPP film, where PDDA is poly(diallyldimethy lammonium chloride, deposited on a linear optical fiber, which was reported in a previous study [22]. In addition, when considering the molecular weights of ammonia (17.03 g mol^{-1}) and TMA (59.11 g mol^{-1}), the LOD for ammonia was about 17 times smaller than that for TMA in the gas phase, which implies a higher selectivity of the sensor to ammonia.

Figure 6a,b show the dynamic responses of the intensity changes and the intensity change-based relative SRs measured at 706 nm for ammonia, amines, and non-amine analytes. Non-amines such as hexanol, ethanol, propanol, acetone, and methanol exhibited slightly decreased intensity changes compared to those of ammonia and other amines. It was confirmed that the transmission spectra of the U-bent fiber sensor were influenced over the entire spectral range (Figure S6). These changes in the transmission spectra may be attributed to increased light scattering due to the RI values of the non-amine analytes [32], because the order of RI for the non-amine analytes is hexanol (1.4178) > propanol (1.377) > acetone (1.36) ≈ ethanol (1.36) > methanol (1.33) ≈ water (1.33). However, the effects of the RI cannot be generalized, such as in the cases of toluene (1.50) and Py (1.51), where the intensity remained unchanged at less than 2 mV over the tested concentration range of ~10,000 and 100 ppm, respectively, as shown in Figure S7. Additionally, the film structure may be another reason. The film includes phenyl and secondary amine moieties, which increase the optical thickness and result in a decrease in the intensity over the entire spectral range when the film is exposed to alcoholic and polar analytes. Therefore, we conclude that the DAR/TSPP film with PSS is specifically sensitive and size-selective to ammonia among the low-molecular-weight volatile amine analytes through acid–base interactions.

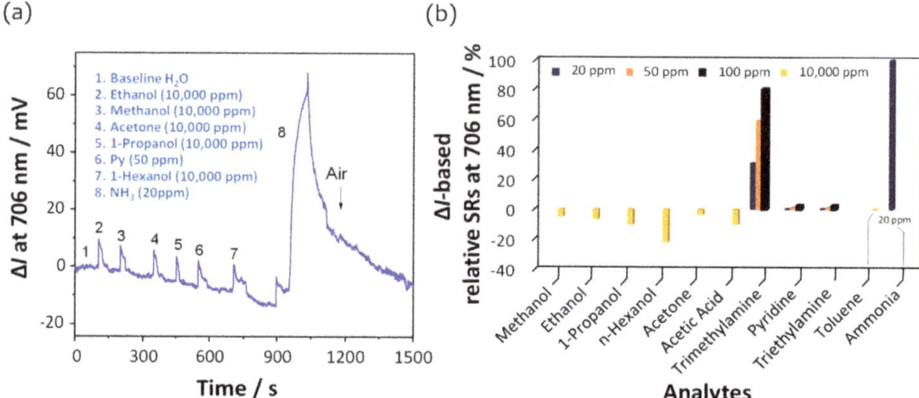

Figure 6. (**a**) Dynamic response to ammonia and other chemical analytes. (**b**) Comparison of the intensity change-based relative SRs at 706 nm on exposure to ammonia (20 ppm), TMA, TEA, Py (20 ppm, 50 ppm, 100 ppm, respectively), and all other analytes at 10,000 ppm.

3.6. Stability and Reproducibility of the Sensing System

To confirm the long-term stability of the SR, the DAR/TSPP+PSS (0.025 wt%) film was repeatedly exposed to 7 ppm (sol) ammonia over seven days. As shown in Figure 7a, the ammonia SR was almost stable over seven days, showing an average intensity change of 34.8 ± 1.5 mV, which is approximately 77% of the first measurement result of the as-prepared film, where the standard deviation of ±1.5 mV represents a variation of approximately 0.4 ppm in ammonia concentration from the sensitivity value (4.05 mV ppm^{-1}) in Table 2. As mentioned above, this small decrease in intensity at 706 nm may be due to the optimal placement of non-crosslinked TSPP molecules inside the film. In addition, similar results were obtained in a series of repeated ammonia exposure experiments over a short period of time (inset in Figure 7a).

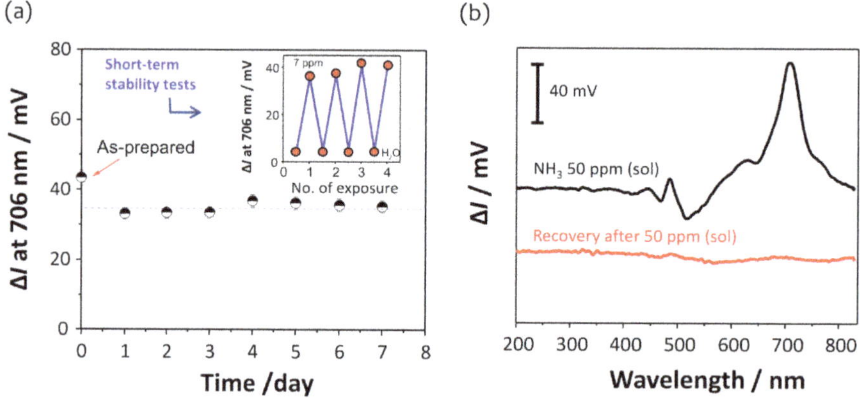

Figure 7. (**a**) Long-term stability of the sensor response due to the repeated exposure (n = 5) to 7 ppm (sol) ammonia over seven days. The inset shows repeatable switching of intensity changes obtained in a series of repeated ammonia exposure experiments in a short period of time. (**b**) Intensity changes in the transmission spectra due to the exposure of the U-bent OFS to 50 ppm (sol) ammonia and after flushing with dry air.

The stability and reproducibility of the sensor response was also confirmed when the sensor film was exposed to 50 ppm (sol) ammonia (Figure 7b). Improvements in sensor response and recovery times were observed in the PSS-embedded film when a 50 ppm (sol) ammonia was repeatedly applied to both sensor films with and without PSS (0.025%) (Figure S8). A twice as long exposure duration (>120 s) was required for the sensor film without PSS compared to that (ca.60 s) of the film containing PSS. Additionally, a much longer time was required for the recovery of the sensor response. One interesting finding was that short-time exposure (<30 s) to ammonia gas resulted in high stability and quick recovery of the sensor response, as reported in a previous study [33]. Herein, the sensor response to a relatively high ammonia concentration (50 ppm) even for a long exposure duration was fully reversible and repeatable over the entire spectral range, as shown in Figure 7b, which was achieved by flushing the film with dry air for a few seconds.

3.7. Film Structure and Morphology

Figure 8 shows the SEM images of the DAR/TSPP and DAR/TSPP PSS (0.025 wt%) sensor films before (Figure 8a,c) and after (Figure 8b,d) ammonia gas exposure. Both films showed rod-like unique structures before gas exposure. Particularly, the DAR/TSPP+PSS (0.025 wt%) film exhibited more extended aggregates and more uniform particles (Figure 8c). However, the film morphologies drastically changed and most of the aggregated structures disappeared after exposure to ammonia gas. On comparing the magnified SEM images in Figure 8e,f, dozens of holes were observed after ammonia gas exposure in both sensor films, particularly in the film without PSS. These differences in film morphology upon the introduction of a small amount of PSS result in the high-performance ammonia gas sensing ability of the DAR/TSPP+PSS (0.025 wt%) film. As morphological changes occur in the film after gas exposure, the light-absorbing and scattering abilities of the exposed film vary compared to those without gaseous exposure, resulting in a change in the evanescent wave and output light intensities of the U-bent OFS.

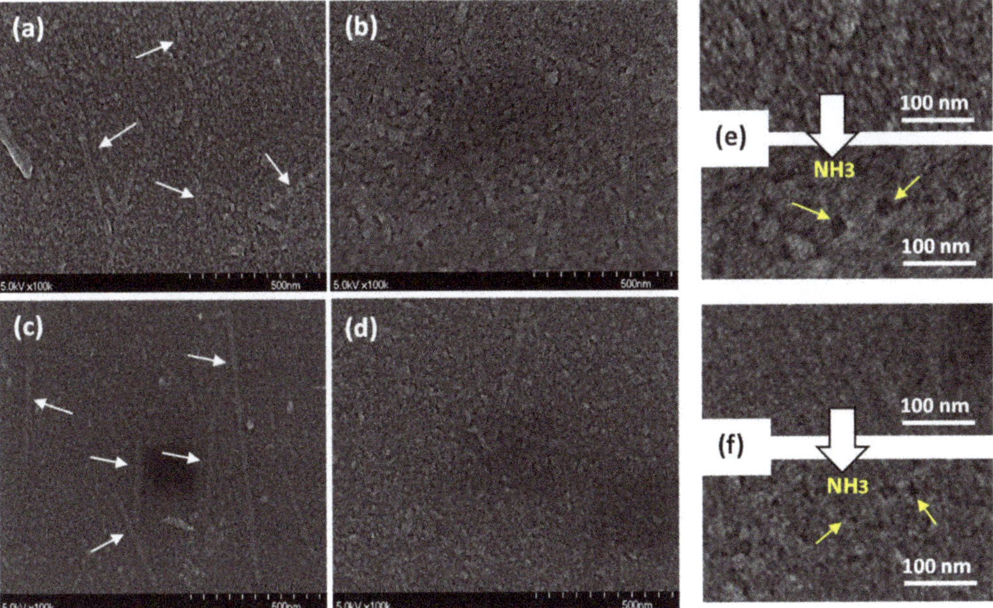

Figure 8. SEM images of (**a**,**b**) 5-cycle DAR/TSPP and (c and d) 5-cycle DAR/TSPP+PSS (0.025 wt%) films deposited on silicon wafer substrates (a and c) before exposure and (**b**,**d**) after exposure to 10,000 ppm (sol) ammonia gas for 2 min. The arrows in white in (**a**,**c**) point to TSPP J-aggregates. Magnified SEM images showing dozens of holes (see arrows in yellow in both cases) after ammonia gas exposure in the sensor films (**e**) without PSS and (**f**) with PSS (0.025 wt%).

The formation of covalent linkages following the decomposition of the diazonium group was further verified by FT-IR measurements. Two absorption peaks observed at 2168 and 2222 cm^{-1}, which originate from the symmetric and asymmetric stretching vibrational modes of the diazonium ion (N_2^+), respectively [23], disappeared entirely after UV irradiation, indicating the decomposition of the diazonium groups. Details of other FT-IR absorption peaks are shown in Figure S9.

3.8. Sensing Mechanism

The exposure of the sensor films to ammonia gas deprotonated the J-aggregated TSPP and transformed it to its monomeric state. This deprotonation process distorts and reduces the TSPP J-aggregates, decreasing the absorbance of the film at 706 nm, thereby causing an increase in the absorbance at 435 nm, which indicates the partial transformation to the mono-acidic state. The mono-acidic TSPP molecules were protonated again through ammonia removal from the film after flushing with dry air. As a result, the TSPP J-aggregates were reformed inside the film, regenerating the baseline.

The sensor fabrication and sensing mechanism of the photocrosslinked DAR/TSPP sensor films with and without PSS are shown in Figure 9. The coexistence of PSS in DAR/TSPP layered films can improve the sensitivity and reproducibility of the sensor response owing to the following two factors. First, the enhanced electrostatic interaction between the DAR and PSS layers increases the number of free TSPP molecules. This facilitates reversible conversion between their deprotonated and protonated forms. Second, there was less condensation on the films. The water molecules adsorbed on the film facilitate the formation of a long residue of ammonium ions (NH_4^+) produced by the deprotonation of TSPP, requiring a longer recovery time for the sensor response. However, as illustrated in Figure 9b–d, film formation with excess PSS prevents the introduction of TSPP J-aggregates in the film, resulting in a reduced number of ammonia binding sites. The further enhanced film rigidity also reduces the flexibility of the TSPP aggregates inside the film and impedes gas diffusion into the film matrix.

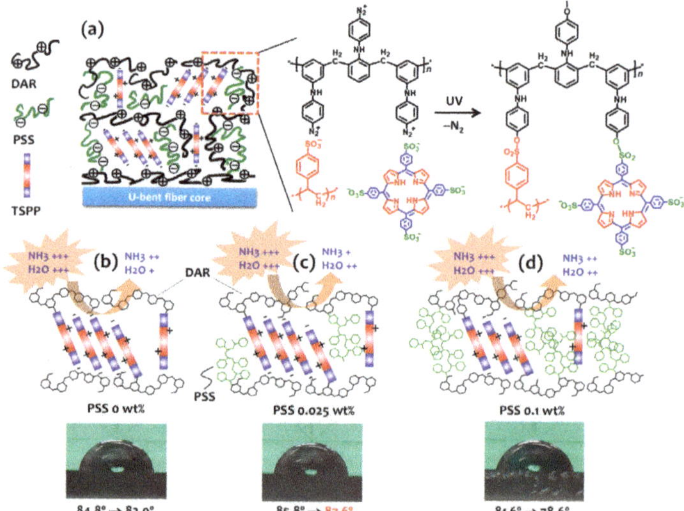

Figure 9. (**a**) Photochemical crosslinking reaction between DAR and the mixture of TSPP and PSS within the multilayered thin film. Schematic illustration of the DAR/TSPP+PSS films with (**b**) PSS 0 wt%, (**c**) PSS 0.025 wt%, and (**d**) PSS 0.1 wt% and photographs of contact angles of the corresponding films. The arrows indicate changes in the contact angle from DI water to a 70-ppm ammonia solution.

The contact angles of the sensor films may help understand the changes in hydrophobicity due to the PSS content employed in the films. Contrary to our expectations, all sensor films showed similar contact angles of over 80° after photocrosslinking by UV irradiation. Compared to the low contact angles of typical layered films prepared via electrostatic self-assembly [34], the DAR-based covalently crosslinked sensor films have higher hydrophobicity. Interestingly, the contact angle of the DAR/TSPP+PSS (0.025 wt%) film changed from 85.8° to 87.6° in the presence of 70 ppm (sol) ammonia after pure water, despite the decrease in the contact angle in the other two films (0 wt% and 0.1 wt%). This result suggests that adding an appropriate amount of PSS (e.g., 0.025 wt% for 1 mM TSPP in water) encouraged rapid gas diffusion due to the improved hydrophobicity inside the film, resulting in momentary ammonia gas condensation at the solid–gas interface.

As a summary, based on the above results, the key features of the U-bent OFS proposed in this study are compared with the previously reported LBL OFSs for ammonia sensing, as listed in Table 3. The PSS-assisted proposed sensor fabrication has the following advantages: (i) long-term stability over seven days, owing to the covalently crosslinked LBL film attachment, (ii) less leakage or detachment of the indicator (TSPP) in a wet atmosphere, (iii) improved hydrophobicity inside the film structure, (iv) efficient transport of the ammonia analyte to the sensing layer, and (v) enhanced evanescent wave due to the U-bent optical fiber geometry. To date, to the best of our knowledge, there is no sensor with satisfactory sensitivity and selectivity for the detection of ammonia from highly humidified samples.

Table 3. A summary based on the comparison of the proposed ammonia sensor to the previously published OFSs.

Sensing Platform	Sensing Material	Fabrication Method	LOD (ppm)	Response Time	Recovery Time	Humidity Range
U-bent optical fiber (this study)	DAR/TSPP+PSS	Crosslinked LbL	0.03	30 s (r.t.)	30 s (r.t.)	85%
Fiber tip [35]	TSPP	Sol–gel	0.15	83 s	NA	<70%
U-bent optical fiber [36]	Bromocresol purple	Dip coating sol–gel	10 (55.5 °C)	5 min (r.t.) 10 s (55.5 °C)	20 min (r.t.) 10 min (55.5 °C)	NA
Side polished fiber [37]	Graphene/polyaniline	Chemical in-situ polymerization	22.5	112 s	185 s	NA
Linear optical fiber [22]	TSPP/PDDA	LbL	3	96 s	192 s	NA
Fiber optic grating [28]	TSPP/PDDA	LbL	0.67	NA	NA	NA
Tapered fiber [38]	TSPP/PAH *	LbL	2	100 s	240 s	NA

NA: not applicable; r.t.: room temperature; * poly(allylamine hydrochloride).

4. Conclusions

Photocrosslinking of DAR and TSPP or DAR and a binary mixture of TSPP and PSS on the core of a U-bent optical fiber was demonstrated for the fabrication of a humidity-resistant OFS. The use of PSS in the film helped obtain stable sensor responses with good gas diffusion into the film. However, the addition of excess PSS inhibited the flexibility of the film and was not beneficial for ammonia gas sensing. Thus, in this study, the optimum PSS concentration was 0.025 wt% with 1 mM TSPP in water. The optimized U-bent OFS exhibited optical responses that had a linear relationship with the aqueous ammonia concentration in the range of 0–17 ppm. The effective response time was less than 30 s, and the baseline was quickly regenerated by flushing the film with dry air after each test. The current sensor system enables the detection of ammonia gas at approximately 30 ppb (parts per billion) concentration. So far, a few standard methods have been developed for ammonia gas sensing, mainly focused on solid-state or electrochemical sensors, along with limited optical sensors. More advanced methods such as gas chromatography–mass spectrometry (GC–MS) and selective ion flow tube–mass spectrometry (SIFT–MS) methods enable very accurate ammonia measurement. However, their use still has some drawbacks in terms of practicality, such as analysis by qualified staff, time-consuming sample preparation and measurement, and complicated operation and maintenance of equipment [39].

Humidity is one of the most influential sources of interference in the development of chemical sensors [40] and no practical method has been demonstrated to solve it yet. From this perspective, the sensor system demonstrated in this study offers a potential and feasible application of chemical sensors to detect ammonia in human breath and in industrial samples as well as ammonia mixed with other amine gases at almost saturated humidity levels. Furthermore, the current study provides a methodology for developing sensor architecture capable of utilizing a non-conventional photocrosslinking LBL method. Our future work will focus on improving the current sensor system by precisely investigating the contribution of hydrophobicity applied in the sensor films.

Supplementary Materials: The following are available online at https://www.mdpi.com/article/10.3390/s21186176/s1, Figure S1: Calibration curves between the actual amine gas concentrations and the corresponding amine concentrations in the solutions, Figure S2: Relative humidity changes inside the measurement chamber during the sample measurement, Figure S3: Evolution of the transmission spectra of the DAR/TSPP+PSS (0.025 wt%) alternate layers deposited onto the 1-cm-long stripped core of a U-bent optical fiber, Figure S4: Comparison of the SRs and calibration curves of the OFS coated with a DAR/TSPP+PSS (0.025%) film when exposed to ammonia gas for 30 s and 120 s, Figure S5: A baseline of the DAR/TSPP+PSS (0.025 wt%) film in a steady state before exposure to ammonia, Figure S6: Intensity changes in the transmission spectra due to the exposure of the U-bent OFS coated with a DAR/TSPP+PSS (0.025%) to different non-amine analytes, Figure S7: Intensity changes in the transmission spectra due to the exposure of the U-bent OFS coated with a DAR/TSPP+PSS (0.025%) to toluene (10,000 ppm) and Py (100 ppm), Figure S8: Comparison of dynamic SRs at 706 nm upon repeated exposure to 50 ppm (sol) of ammonia for the 5-cycle DAR/TSPP+PSS (0 and 0.025 wt%) films, Figure S9: FT-IR spectra of a 10-cycle DAR/TSPP+PSS (0.025 wt%) film deposited on a gold-coated silicon wafer substrate before and after UV irradiation.

Author Contributions: S.A. performed the experiment, collected the data, analyzed the data, and contributed to the writing of the initial draft. Y.P. performed the experiment for surface analysis. H.O. and S.O. performed the experiment for DAR synthesis. S.K. helped analyze the data and revised the manuscript. S.-W.L. conceived the idea of the study, helped perform the analysis with constructive discussions, and revised the manuscript for finalizing of this paper. All authors have read and agreed to the published version of the manuscript.

Funding: This work was funded by the EPSRC Cyclops Healthcare Network (EP/N026985/1). S.-W. Lee acknowledges the grant-in-aid for scientific research (C) (20K09223) from the Japan Society for the Promotion of Science.

Institutional Review Board Statement: Not applicable.

Informed Consent Statement: Not applicable.

Data Availability Statement: Not applicable.

Acknowledgments: We kindly acknowledge the supports from Shinkouseiki Co. Ltd. and Toyoko Co. Ltd. as research partners for this project.

Conflicts of Interest: The authors declare no conflict of interest.

References

1. Pospíšilová, M.; Kuncová, G.; Trögl, J. Fiber-optic chemical sensors and fiber-optic bio-sensors. *Sensors* **2015**, *15*, 25208–25259. [CrossRef]
2. Liu, L.; Marques, L.; Correia, R.; Morgan, S.P.; Lee, S.-W.; Tighe, P.; Fairclough, L.; Korposh, S. Highly sensitive label-free antibody detection using a long period fibre grating sensor. *Sens. Actuators B Chem.* **2018**, *271*, 24–32. [CrossRef]
3. Kharaz, A.; Jones, B.E. A distributed optical-fibre sensing system for multi-point humidity measurement. *Sens. Actuators A Phys.* **1995**, *47*, 491–493. [CrossRef]
4. Rajamani, A.S.; Murugan, D.; Sai, V.V.R. Plastic fiber optic sensor for continuous liquid level monitoring. *Sens. Actuators A Phys.* **2019**, *296*, 192–199. [CrossRef]
5. Cennamo, N.; Massarotti, D.; Conte, L.; Zeni, L. Low cost sensors based on SPR in a plastic optical fiber for biosensor implementation. *Sensors* **2011**, *11*, 11752–11760. [CrossRef]
6. Lomer, M.; Quintela, A.; López-Amo, M.; Zubia, J.; López-Higuera, J.M. A quasi-distributed level sensor based on a bent side-polished plastic optical fibre cable. *Meas. Sci. Technol.* **2007**, *18*, 2261–2267. [CrossRef]
7. Feng, D.-J.; Zhang, M.-S.; Liu, G.; Liu, X.-L.; Jia, D.-F. D-Shaped plastic optical fiber sensor for testing refractive index. *IEEE Sens. J.* **2014**, *14*, 1673–1676.
8. Satija, J.; Punjabi, N.S.; Sai, V.V.R. Optimal design for U-bent fiber-optic LSPR sensor probes. *Plasmonics* **2014**, *9*, 251–260. [CrossRef]
9. Timmer, B.; Olthuis, W.; Berg, A. Ammonia sensors and their applications—A review. *Sens. Actuators B Chem.* **2005**, *107*, 666–677. [CrossRef]
10. Occupational Safety and Health Administration. Available online: https://www.osha.gov/laws-regs/regulations/standardnumber/1910/1910.1000TABLEZ1 (accessed on 8 May 2021).
11. Shipman, M.A.; Symes, M.D. Recent progress towards the electro-synthesis of ammonia from sustainable resources. *Catal. Today* **2017**, *286*, 57–68. [CrossRef]
12. Liu, L.; Fei, T.; Guan, X.; Zhao, H.; Zhang, T. Humidity-activated ammonia sensor with excellent selectivity for exhaled breath analysis. *Sens. Actuators B Chem.* **2021**, *334*, 129625. [CrossRef]
13. Shahmoradi, A.; Hosseini, A.; Akbarinejad, A.; Alizadeh, N. Noninvasive detection of ammonia in the breath of hemodialysis patients using a highly sensitive ammonia sensor based on a polypyrrole/sulfonated graphene nanocomposite. *Anal. Chem.* **2021**, *93*, 6706–6714. [CrossRef]
14. Davies, S.; Spanel, P.; Smith, D. Quantitative analysis of ammonia on the breath of patients in end-stage renal failure. *Kidney Int.* **1997**, *52*, 223–228. [CrossRef] [PubMed]
15. Narasimhan, L.R.; Goodman, W.; Patel, C.K.N. Correlation of breath ammonia with blood urea nitrogen and creatinine during hemodialysis. *Proc. Natl. Acad. Sci. USA* **2001**, *98*, 4617–4621. [CrossRef] [PubMed]
16. Ong, J.P.; Aggarwal, A.; Krieger, D.; Easley, K.A.; Karafa, M.T.; Lente, F.; Arroliga, A.C.; Mullen, K.D. Correlation between ammonia levels and the severity of hepatic encephalopathy. *Am. J. Med.* **2003**, *114*, 188–193. [CrossRef]
17. Kusters, J.G.; Vliet, A.H.M.; Kuipers, E.J. Pathogenesis of helicobacter pylori infection. *Clin. Microbiol. Rev.* **2006**, *19*, 449–490. [CrossRef]
18. Bannov, A.G.; Popov, M.V.; Brester, A.E.; Kurmashov, P.B. Recent advances in ammonia gas sensors based on carbon nanomaterials. *Micromachines* **2021**, *12*, 186. [CrossRef]
19. Rodríguez, A.; Zamarreño, C.; Matías, I.; Arregui, F.; Cruz, R.; May-Arrioja, D. A fiber optic ammonia sensor using a universal pH indicator. *Sensors* **2014**, *14*, 4060–4073. [CrossRef] [PubMed]
20. Korposh, S.; Kodaira, S.; Lee, S.-W.; Batty, W.J.; James, S.W. Nano-assembled thin film gas sensor: II. An intrinsic high sensitive fibre optic sensor for ammonia detection. *Sens. Mater.* **2009**, *21*, 179–189. [CrossRef]
21. Korposh, S.; Selyanchyn, R.; Yasukochi, W.; Lee, S.-W.; James, S.W.; Tatam, R.P. Optical fibre long period grating with a nanoporous coating formed from silica nanoparticles for ammonia sensing in water. *Mater. Chem. Phys.* **2012**, *133*, 784–792. [CrossRef]
22. Korposh, S.; Kodaira, S.; Selyanchyn, R.; Ledezma, F.H.; James, S.W.; Lee, S.-W. Porphyrin-nanoassembled fiber-optic gas sensor fabrication: Optimization of parameters for sensitive ammonia gas detection. *Opt. Laser Technol.* **2018**, *101*, 1–10. [CrossRef]
23. Sun, J.; Wu, T.; Liu, F.; Wang, Z.; Zhang, X.; Shen, J. Covalently attached multilayer assemblies by sequential adsorption of polycationic diazo-resins and polyanionic poly(acrylic acid). *Langmuir* **2000**, *16*, 4620–4624. [CrossRef]
24. Sun, J.; Wu, T.; Zou, B.; Zhang, X.; Shen, J. Stable entrapment of small molecules bearing sulfonate groups in multilayer assemblies. *Langmuir* **2001**, *17*, 4035–4041. [CrossRef]
25. Cao, W.X.; Ye, S.J.; Cao, S.G.; Zhao, C. Novel polyelectrolyte complexes based on diazo-resins. *Macromol. Rapid Commun.* **1997**, *18*, 983–989. [CrossRef]
26. Zhao, C.; Chen, J.Y.; Cao, W.X. Synthesis and characterization of diphenylamine diazonium salts and diazoresins. *Angew. Makromol. Chem.* **1998**, *259*, 77–82. [CrossRef]
27. Wang, T.; Korposh, S.; James, S.W.; Tatam, R.P.; Lee, S.-W. Optical fiber long period grating sensor with a polyelectrolyte alternate thin film for gas sensing of amine odors. *Sens. Actuators B Chem.* **2013**, *185*, 117–124. [CrossRef]
28. Wang, T.; Yasukochi, W.; Korposh, S.; James, S.W.; Tatam, R.P.; Lee, S.-W. A long period grating optical fiber sensor with nano-assembled porphyrin layers for detecting ammonia gas. *Sens. Actuators B Chem.* **2016**, *228*, 573–580. [CrossRef]

29. Takagi, S.; Eguchi, M.; Tryk, D.A.; Inoue, H. Porphyrin photochemistry in inorganic/organic hybrid materials: Clays, layered semiconductors, nanotubes, and mesoporous materials. *J. Photochem. Photobiol. C* **2006**, *7*, 104–126. [CrossRef]
30. Zha, S.; Li, X.; Yang, M.; Sun, C. Fabrication and characterization of covalently attached multilayer films containing iron phthalocyanine and diazo-resins. *J. Mater. Chem.* **2004**, *14*, 840–844. [CrossRef]
31. Gregory van Patten, P.; Shreve, A.P.; Donohoe, R.J. Structural and photophysical properties of water-soluble porphyrin associated with polycations in solution and electrostatically-assembled ultrathin films. *J. Phys. Chem. B* **2000**, *10*, 5986–5992. [CrossRef]
32. John, A.R.; William, B.B.; Theodore, K.S. *Organic Solvents Physical Properties and Methods of Purification: Techniques of Chemistry*, 4th ed.; Wiley: New York, NY, USA, 1986.
33. Lee, S.-W.; Takahara, N.; Korposh, S.; Yang, D.-H.; Toko, K.; Kunitake, T. Nano-assembled thin film gas sensor. III. Sensitive detection of amine odors using TiO_2/polyacrylic acid ultrathin film QCM sensors. *Anal. Chem.* **2010**, *82*, 2228–2236. [CrossRef]
34. Smith, A.R.; Ruggles, J.L.; Yu, A.; Gentle, I.R. Multilayer nanostructured porphyrin arrays constructed by layer-by-layer self-assembly. *Langmuir* **2009**, *25*, 9873–9878. [CrossRef] [PubMed]
35. Liu, L.; Morgan, S.P.; Correia, R.; Lee, S.-W.; Korposh, S. Multi-parameter optical fiber sensing of gaseous ammonia and carbon dioxide. *J. Light. Technol.* **2020**, *38*, 2037–2045. [CrossRef]
36. Cao, W.; Duan, Y. Optical fiber-based evanescent ammonia sensor. *Sens. Actuators B Chem.* **2005**, *110*, 252–259. [CrossRef]
37. Khalaf, A.L.; Mohamad, F.S.; Abdul Rahman, N.; Lim, H.N.; Paiman, S.; Yusof, N.A.; Mahdi, M.A.; Yaacob, M.H. Room temperature ammonia sensor using side-polished optical fiber coated with graphene/polyaniline nanocomposite. *Opt. Mater. Exp.* **2017**, *7*, 1858–1870. [CrossRef]
38. Jarzebinska, R.; Korposh, S.; James, S.; Batty, W.; Tatam, R.; Lee, S.-W. Optical gas sensor fabrication based on porphyrin-anchored electrostatic self-Assembly onto tapered optical fibers. *Anal. Lett.* **2012**, *45*, 1297–1309. [CrossRef]
39. Bielecki, Z.; Stacewicz, T.; Smulko, J.; Wojtas, J. Ammonia gas sensors: Comparison of solid-state and optical methods. *Appl. Sci.* **2020**, *10*, 5111. [CrossRef]
40. Shiba, S.; Yamada, K.; Matsuguchi, M. Humidity-resistive optical NO gas sensor devices based on cobalt tetraphenylporphyrin dispersed in hydrophobic polymer matrix. *Sensors* **2020**, *20*, 1295. [CrossRef]

Article

Hg^{2+} Optical Fiber Sensor Based on LSPR Generated by Gold Nanoparticles Embedded in LBL Nano-Assembled Coatings

María Elena Martínez-Hernández [1,*], Javier Goicoechea [1,2] and Francisco J. Arregui [1,2]

1 Department of Electrical, Electronic and Communication Engineering, Universidad Publica de Navarra, Edif. Los Tejos, Campus Arrosadía, 31006 Pamplona, Spain; javier.goico@unavarra.es (J.G.); parregui@unavarra.es (F.J.A.)

2 Institute of Smart Cities (ISC), Universidad Publica de Navarra, Campus Arrosadia, 31006 Pamplona, Spain

* Correspondence: mariaelena.martinez@unavarra.es

Received: 15 October 2019; Accepted: 7 November 2019; Published: 10 November 2019

Abstract: Mercury is an important contaminant since it is accumulated in the body of living beings, and very small concentrations are very dangerous in the long term. This paper reports the fabrication of a highly sensitive fiber optic sensor using the layer-by-layer nano-assembly technique with gold nanoparticles (AuNPs). The gold nanoparticles were obtained via a water-based synthesis route that use poly acrylic acid (PAA) as stabilizing agent, in the presence of a borane dimethylamine complex (DMAB) as reducing agent, giving PAA-capped AuNPs. The sensing mechanism is based on the alteration of the Localized Surface Plasmon Resonances (LSPR) generated by AuNPs thanks to the strong chemical affinity of metallic mercury towards gold, which lead to amalgam alloys.

Keywords: fiber optic sensor; gold nanoparticles; localized surface plasmon resonance; mercury; ppb

---

1. Introduction

The presence of mercury in the environment is a real concern nowadays. It is well known that mercury not only causes damage to the environment, but also to human health [1]. It is a highly toxic element known to cause DNA damage, lipid peroxidation, and protein oxidation and deactivation and is also associated with cardiovascular diseases [2]. This has become a priority matter in the European Union (EU) and the United States Environmental Protection Agency (US-EPA) [3], which seek to take actions against diverse harmful agents that attack the environment like solvents, hydrocarbons, pesticides, and heavy ions (mercury among them). Their objective is to improve the determination of human exposure through integrated monitoring of the environment and food [4,5]. In 2006, the International Conference on Chemicals Management adopted the "Dubai Declaration on International Chemicals Management", the "Overarching Policy Strategy", and endorsed the "Global Plan of Action", in which priority attention is given to mercury [6].

Those international institutions have regulated that any water source and aquifer as well as some specific food products should be monitored and controlled in order to guarantee that they have admissible levels of a series of dangerous contaminants [7], but this task cannot be done nowadays because these tests would require unaffordable costs and complexity. That is why there are many research works that are focused on finding simpler, better, and more accurate ways to detect mercury, where the biggest challenge is to obtain quick, cost-effective and accessible results.

To solve this problem, new methods and perspectives with the use of different sensor devices have been reported. Classical approaches include the monitoring of electrochemical reactions, using techniques like galvano-static techniques, impedance measurement, electrochemiluminescence, and others [1–3]. However, most suffer some reproducibility and stability problems [4].

Among the different sensing materials, gold nanoparticles are one of the most interesting materials for optical sensing applications [5], because of their stability, compatibility with the aqueous medium, easy surface functionalization along with miniaturization [6], and their optical properties. When gold nanoparticles interact with light, there is a resonant light-matter energy coupling known as Localized Surface Plasmon Resonances (LSPR) which can be used as a sensing signal [7–10]. The LSPR occurs thanks to the energy transfer from incident light to certain collective oscillation modes of the free electrons within the nanoparticles that creates an intense optical absorption band. The location of this resonant peak in the visible or infrared region depends on multiple factors such as shape, size, aggregation state, distribution or interaction of the nanoparticles [9]. The consequences of exciting the LSPR are the selective absorption of certain excitation wavelengths and the generation of locally enhanced or amplified electromagnetic fields (EM) on the surface of the nanoparticles and their resonant condition is very sensitive to refractive index variations of the close surrounding medium and the surface chemistry of the nanoparticles [11]. Some studies base their sensing mechanism on the variation of the optical absorbance intensity of the LSPR bands [12–14] of the gold nanoparticles simply due to their surface interaction with mercury ions. Sensors of this type can have low detection limits [15,16]. Furthermore some of the reported works require the use of additional measuring techniques such as ellipsometry [17], SPR reflectometry [18] or involving biological reactions, such as aptamer-based recognition [15], allowing highly sensitivities (LODs around), but increasing the sensors complexity and their cost. Those approaches suppose a limitation for the practical use of such sensors, and the development of more robust, simple and effective sensitive coatings is still a challenge nowadays.

Fiber optic sensors can be a simpler and powerful alternative to these nanoparticles dispersions analysis because have small size, electromagnetic immunity, electrically passivity, and biocompatibility [19]. One of the most common approaches to create optical fiber sensors is the immobilization of the sensitive material onto the surface of the optical fiber. In this manner, the guided light is altered by the interaction with the sensitive material whose optical properties can be affected by the presence of the target to be measured. Therefore, the photonic signal traveling through the fiber will be also modified, which constitutes one of the most common transduction principles of optical fiber sensors for chemical measurements [20]. So far, gold nanoparticles are the most popular solution for the development of highly-sensitive mercury fiber optic sensors, thanks to their stability, small size, low cost, and outstanding optical properties. In most of the approaches, additional molecules or biomolecules are needed to cause this LSPR variation, such as the tendency of mercury to form complexes with certain proteins [13] or the use IgG–anti IgG as bioreceptor–analyte pair [14]. Other sensors study the change of LSPR resonance wavelength. For example, it has been reported the plasmon-coupling effect in gold nanoparticles core-satellites nanostructures linked by thymine(T)-rich DNA hybridization [21]. It is known that the shape and distribution of gold nanoparticles can generate changes in the LSPR, causing wavelength shifts [11]. The process of Au–Hg alloy is able to modify the shape of gold nanoparticles causing changes remarkable blue shifts. Such changes occurred because of the chemical modification of the nanoparticles near their surface (Hg–Au amalgam formation) modify their effective size and shape [22] altering the LSPR resonant condition.

In this work, it is proposed the embedding of gold nanoparticles in a polymeric matrix that allow the interaction with mercury ions (Hg^{2+}). The sensing mechanism is based on the strong chemical affinity of metallic mercury (Hg^0) towards gold [18] to form stable amalgam-like alloys [17], and consequently altering the LSPR resonance of the gold nanoparticles, therefore, providing a wavelength-based sensing signal. It has been already reported that the reaction of metallic mercury on the surface of the gold nanoparticles can cause the change of their shape, affecting to the LSPR resonance conditions [22].

The layer-by-layer nano-assembly technique is used here for such embedding of the metallic nanoparticles in the matrix that can facilitate the gold-mercury interaction. This sensing mechanism is simpler than the previous approaches reported in the literature and does not involve the utilization of auxiliary biomolecules with the gold nanoparticles.

2. Materials and Methods

2.1. Materials

The polymer poly (allylamine hydrochloride) (PAH) (Mw~15.000) was used as polycation during the LbL process. For the synthesis of AuNPs it was used poly (acrylic acid) (PAA) 35 wt% solution in water, Borane dimethylamine complex (DMAB) and Gold (III) chloride trihydrate. The pH of the solutions were adjusted using HCl and NaOH. The mercury samples were prepared with Mercury (II) chloride ($HgCl_2$) in buffer phosphate. For the buffer solutions it was used sodium phosphate dibasic (Na_2HPO_4) and sodium phosphate monobasic (NaH_2PO_4). Piranha solution was also used, which is the combination of sulfuric acid (H_2SO_4) with hydrogen peroxide (H_2O_2), 3:1 ratio. All materials were supplied by Sigma Aldrich and aqueous solutions were prepared using ultrapure water with a resistivity of 18.2 MΩ·cm.

2.2. Synthesis Method of the PAA-Capped AuNPs

There are other works that describe different synthesis routes for metallic nanoparticles of various morphologies [23–26]. In this case, to AuNPs synthesis it was used a chemical reduction route carried out in water-based solutions in which the PAA act as a stabilizer [23]. Gold nanoparticles have been prepared by adding 20 mL of $HAuCl_4{\cdot}3H_2O$ (5 mM) to 120 mL of PAA (10 mM). This solution was stirred for 2 h. Afterwards 1 mL of fresh DMAB (0.1 M) solution was added under vigorous stirring, and the reaction was left overnight. All operations were performed at room conditions. UV-VIS absorption spectra of the synthesized nanoparticles dispersions were characterized using a Jasco V-630 spectrophotometer. The UV-VIS absorption spectra of the PAA-AuNP dispersions showed a LSPR absorption band centered at 540 nm. Transmission electron microscopy (TEM) has been used to determine the morphology of the AuNPs, resulting in spherical shape particles, with a diameter ranging from 10 to 20 nm [23].

2.3. Optical Detection Setup

Optical fiber sensors were made from multimode optical fibers 200 μm-core diameter with polymeric cladding, 0.39 NA (THORLABS FT 200EMT). The sensor structure was based on the mechanical removal of the acrylate cladding of a segment of approximately 2 cm of the optical fiber. This removal was performed with the help of a few drops of dry acetone and a blade, exposing the bare optical fiber core, in its entire cylindrical section. Subsequently, this optical fiber segment was immersed for 5 min in piranha solution to eliminate the acetone that could remain. The ends of the optical fiber were terminated using temporary SMA connectors (THORLABS BFT1). The sensor was excited from one of the connectors with a halogen white source and the other end collect the optical response with a CCD spectrometer (HR4000-UV Ocean Optics).

2.4. Layer-By-Layer Nano-Assembly

Using layer-by-layer nano-assembly (LbL) it is possible the deposition of oppositely charged polyelectrolyte ultra-thin layers by dipping the substrates into a sequence of solutions. A solution of PAH (10 mM) was used as polycationic solution, and PAA-capped AuNPs (PAA-AuNPs) dispersion was used as polyanion. The optical fiber substrates were immersed into each charged solution for 5 min. After every polyelectrolyte adsorption step it is necessary to rinse the assembly in ultrapure water with same pH of the polyelectrolytes [24,25]. Each polycation/polyanion layer combination is called bilayer. In this work, a total of six bilayers of (PAH/PAA+AuNPs) are deposited onto the cladding removed optical fiber segment (Figure 1). All solutions were adjusted to pH 7. Before starting the deposition of layer by layer, the entire fiber segment where the cladding was removed was immersed in KOH (1M) for half an hour to achieve substrate surface electrostatic charge

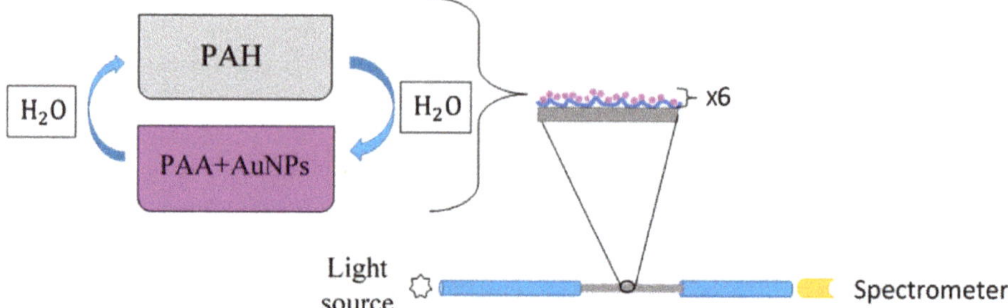

Figure 1. Layer-by-layer nano-assembly built-up of the sensitive coatings. Construction of fiber optic sensor with $(PAH/PAA+AuNPs)_6$ over a cladding-removed 200 μm-core optical fiber.

2.5. Mercury Samples

Phosphate buffer (PB) was prepared dissolving 2.198 g of Na_2HPO_4 in 400 mL of ultrapure water. This solution was stirred for 15 min, then 0.62 g of NaH_2PO_4 was added and 100 mL of ultrapure water, and stirred for 15 min, obtaining a pH = 7.6. The different concentrations of Mercury (II) chloride ($HgCl_2$) were dissolved in the PB. Different concentration mercury samples must be in metallic form to interact with the AuNPs, consequently, before exposing the optical fiber sensor to the mercury ions, it is necessary to reduce Hg^{2+} to Hg^0 using DMAB (12 mL of a freshly prepared DMAB stock solution (0.1 M) as reducing agent. The reaction was stirred at room temperature conditions for 2 h (kept away from direct sunlight). The Hg concentrations analyzed were 1, 2, 4, 8, 10, and 20 ppb. All samples were kept under stirring until the moment of measurement.

In order to vary only the mercury concentration and keep the rest of parameters constant, the DMAB amount is corrected for every sample just adding a certain amount of blank PB stock solution. For every measurement, the optical fiber sensor was immersed into the Buffer PB + DMAB solution prior to the exposure to the mercury ion stock solutions, in order to get a stable baseline for the latter mercury detection.

2.6. Sensors Regeneration

It is known that nitric acid forms highly instable complexes with Hg^{2+} and favors the separation of mercury from gold nanoparticles [26]. The regenerating solution was prepared starting from a stock PB (pH 7.6) and HNO_3 was added dropwise until the pH was lowered to 4.6 and the dissolution was keep at a constant temperature of 55 °C.

2.7. Data Processing

During the immersion in the mercury solutions all spectra were recorded continuously and the LSPR maxima were estimated using a Matlab® algorithm. This provides live information about the time response of the sensors. The results obtained will be estimated by their dynamic response as a way to obtain parameters for rapid estimation before the responses obtained from the sensor.

2.8. Cross-Sensitivity to Other Metals

There are other metals whose presence in the organism is necessary because they are involved in biological functions, however, when they exceed a certain threshold they can be considered toxic, among them we can find zinc and nickel [27]. Consequently it is very important to characterize the mercury sensor cross-sensitivity against other metal ions such as Fe^{2+}, Ni^{2+}, Pb^{2+}, Cd^{2+}, and Zn^{2+}. All the solutions were prepared under the same conditions as the mercury samples. All ionic species for the cross-sensitivity test has been set to the maximum concentration used with the mercury (20 ppb).

This concentration of the other ionic species are significantly higher than the limits required by the regulation [28] like for Fe (6.2 ppb), Cd (3.4 ppb) and Zn (1.8 ppb).

3. Results and Discussion

3.1. Effects of Hg^0 on AuNPs in Dispersion

Prior to the construction of the fiber optic sensor, a preliminary study was made in order to characterize the effects of mercury on the optical properties of the gold nanoparticles in dispersion. It is known the Hg^0 can be bonded onto the surface of Au-based nanomaterials to form a solid amalgam-like alloy [29,30].

Samples in dispersion were analyzed with different Hg concentration and keeping constant the volume and concentration of AuNPs solutions. US-VIS spectra of the dispersions showed a dramatic change of LSPR resonance wavelength clearly seen with the naked eye as a color change (Figure 2).

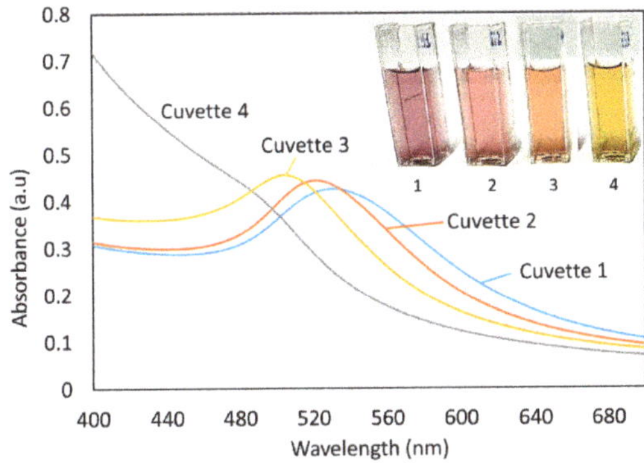

Figure 2. Change of Localized Surface Plasmon Resonances (LSPR) wavelength of PAA-AuNP dispersion with different Hg concentration. Cuvette 1: as prepared only with AuNPs. Cuvette 2: PAA-AuNP + DMAB. Cuvette 3: PAA-AuNP + DMAB and mercury (2.7 ppb). Cuvette 4: PAA-AuNP + DMAB and mercury (27 ppb).

In Figure 2 it is shown the UV-VIS spectrum of cuvette 1 that contain only a PAA-AuNPs dispersion as prepared that shows an violet-reddish color. The spectrum shows the typical LSPR attenuation band centered in 540 nm which is compatible with the synthesis routes available in the literature. Since it is necessary to reduce mercury ions to their metallic form (Hg^0) to enable the amalgam interaction, cuvette 2 is equal to cuvette 1, except that 200 μL of 0.1 M DMAB were added. Here it is observed a slight blue shift of the LSPR resonance wavelength of 8 nm, that remained stable in time. This LSPR variation is probably due to the modification of the polymeric PAA stabilization cap thanks to the interaction with the DMAB. For lower mercury concentrations such as 2.7 ppb (cuvette 3), the displacement also occurs, but to a lesser extent, in 18 nm respect to the cuvette 2 that is the optical reference with no mercury. When 300 μL of Hg (10^{-3} M) were added to the dispersion keeping the same concentration and volume of PAA-AuNPs and DMAB solution, it was obtained 27 ppb of mercury concentration and the LSPR resonance experimented a stronger blue shift, almost disappearing, yielding a clear yellowish color.

In the synthetic process of AuNPs, the reduction of gold ions (Au^{3+}) to gold nanoparticles (Au^0) is possible thanks to the use of a protective agent (PAA), which contributes to control the shape and size of the resultant nanoparticles, preventing their agglomeration or precipitation in the colloidal solution and the DMAB that acts as a reducing agent [9]. The small displacement of LSPR resonance wavelength that occurred in the case of cuvette 1 as a result of interaction between AuNP-PAA with the additional DMAB present in the sample solutions (with no mercury in cuvette 1). This LSPR wavelength shift could be induced by the refractive index variation in the optical fiber immersion media. However, cuvettes 2, 3, and 4 have the same concentrations of AuNP-PAA and DMAB, they only differ in a very small mercury concentration that induces a more severe LSPR resonance displacement thanks to the chemical modification of the AuNPs.

3.2. Obtaining the AuNPs LSPR onto the Fiber Optics

The absorbance of the $(PAH/PAA+AuNPs)_n$ coating onto the optical fiber was registered during the LbL process. With every bilayer increment the absorbance spectrum show an increasing of the intensity around the 540 nm wavelength suggesting a homogeneous growth of the LbL coating. After an optimization study of the LbL process it was found that using six bilayers it is possible to obtain a well-defined LSPR absorption band (Figure 3) consequently this number of bilayers was kept constant for all the sensors in this work.

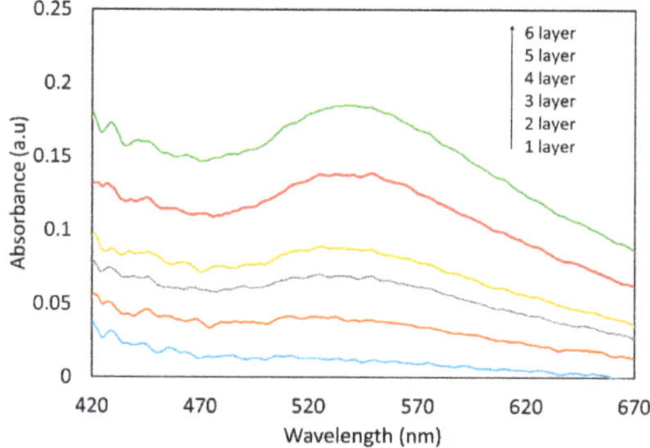

Figure 3. Absorption spectra of LSPR resonance wavelength for every layer of (PAH/PAA+AuNPs) deposited on 200 μm-core optical fiber.

The absorbance of the LbL $(PAH/PAA+AuNPs)_n$ films (being n the number of bilayers) is shown in Figure 3. The absorbance spectra confirm the existence of an absorption band centered at 540 nm, which corresponds to the LSPR of the AuNPs. This demonstrates that the absorption band of the coatings matches with that of the AuNPs dispersion initially synthesized by chemical reduction seen in Figure 2 (cuvette 1).

3.3. Detection of Mercury Ions with Fiber Optic Sensor

In Figure 4, it is shown an initial immersion of the sensor in the Buffer PB + DMAB solution, it was registered a small displacement of LSPR resonance wavelength (approximately 5 nm) and after a few minutes it remained stable. In this work, all sensors were kept in this solution for one hour in order to have a stable baseline for the later mercury detection stage. Nevertheless, shorter immersion times could be also acceptable. After the sensor it was immersed in a 20 ppb mercury sample and there was a variation of absorption with respect to the condition of the baseline. In addition, for the mercury concentration of 20 ppb there is a change in the LSPR resonance wavelength of 15 nm with respect to the Buffer PB + DMAB solution.

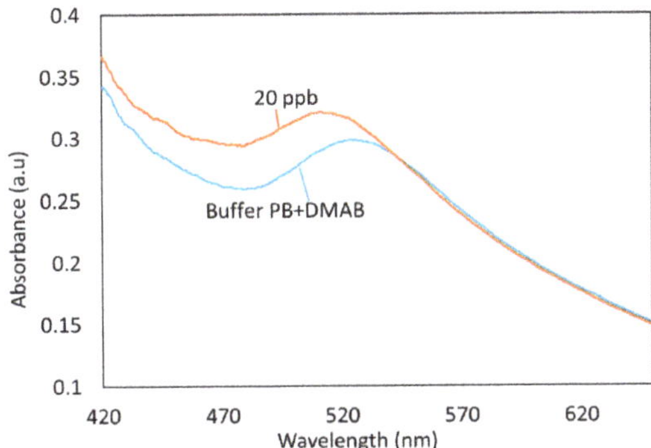

Figure 4. Displacement in wavelength of the LSPR for 20 ppb of mercury concentration.

Different sensors were fabricated with the same materials and methods mentioned in Section 2, and each one was used to detect a particular mercury concentration (shown in Figure 5). Although the wavelength shift of the LSPR band was easily visible in a few seconds for the highest mercury concentration, the $(PAH/PAA+AuNPs)_6$ sensors showed a settling time of nearly 3000 s (from 10% to 90%). Consequently, all the sensors were immersed in the mercury solution for 50 min. All sensors' LSPR bands experimented a blue-shift when exposed to mercury. The absolute wavelength shift increases with the mercury concentration; for 20 ppb of mercury solution, the LSPR maximum wavelength change is 16 nm, and for 1 ppb is 1.11 nm.

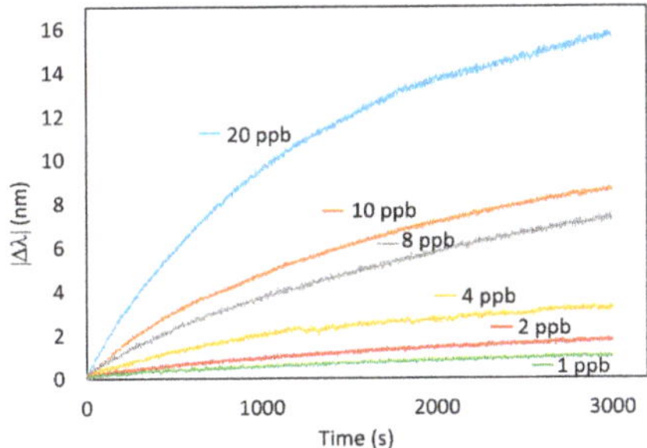

Figure 5. Dynamic response of the sensors to different Hg concentration, ranging from 20 ppb to 1 ppb.

This wavelength-based response can be seen in Figure 6, where it is shown the maximum variation in wavelength $|\Delta\lambda max|$ for each mercury concentration and the linear fitting.

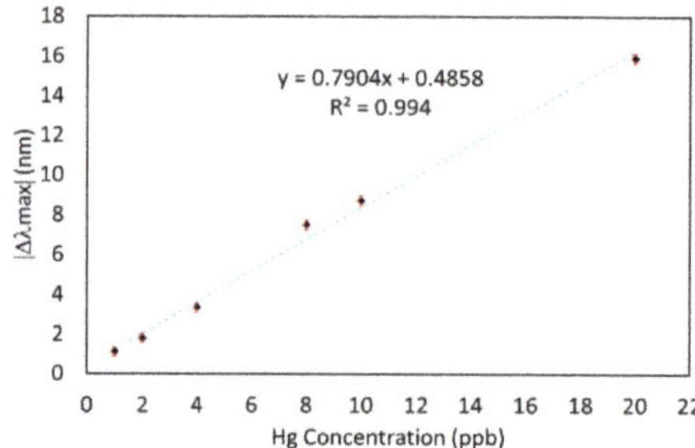

Figure 6. Maximum variation in wavelength (Δλmax) for each mercury concentration (1, 2, 4, 8, 10, and 20 ppb) after 3000 s.

From the continuous data acquisition of the baseline during the immersion in Buffer PB + DMAB (0 ppb), it is possible to calculate the standard deviation (σ). The limit of detection (LOD) of the sensor can be estimated as 3σ, that is 0.147 nm, which is equivalent to 0.7 ppb, which is below the 2 ppb detection limit stated by the US-EPA and 1 ppb for the EU [29,30].

The results presented so far are accurate enough to provide reliable measurements of aqueous samples without any further chemical or biological agent, and they could be performed in the field. Nevertheless, the sensors still need relatively long time measurements. In order to overcome this, a measurement technique is proposed to obtain faster measurements. In this sense, the values of the slopes of each dynamic curve (Figure 5) could be used as a fast estimation parameter. In Figure 7, the slope of the sensor's response approximated by the linear fitting of the first 500 s (roughly 8 min) is plotted for every Hg concentration.

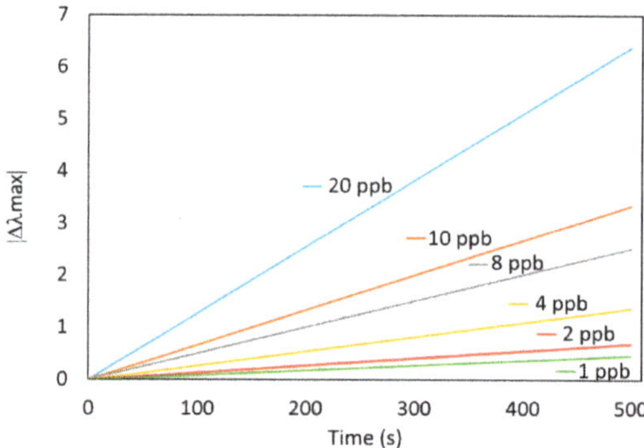

Figure 7. Fast estimation of the optical fiber sensor response using the slope of the linear fitting of the samples of the first 500 s. The Hg concentration has been varied from 1 to 20 ppb.

As can be observed, the value of the slope increases with the increase in mercury concentration, getting lineal response (Figure 8a). In Figure 8b it is shown the high correlation between the absolute wavelength shift and the slope of each curve, meaning that it can be reliably used as a fast response estimator. These results allow estimating the behavior of the sensor in different mercury concentrations in a faster way, after 500 s.

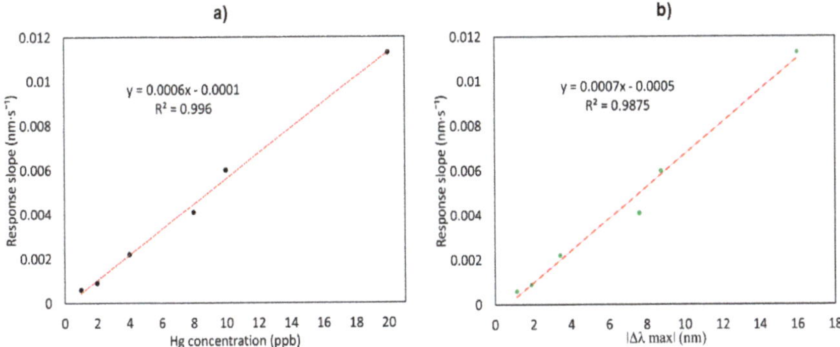

Figure 8. (**a**) Slopes of the linear fitting of the first 500 s vs. mercury concentration. (**b**) Correlation between the mercury measurement using the absolute wavelength shift of the LSPR, and the fast slope estimation.

3.4. Sensor Regeneration

Another critical aspect is the reusability of a single optical fiber sensor for multiple measurements. In fact, it is possible to regenerate the sensor in a HNO_3 solution, the sensor was deposited in the solution mentioned in Section 2.6 for 1 h. During the immersion in the regenerating solution it was observed a red shift of the LSPR resonance wavelength, which is a similar to the first reaction curve. After regeneration the sensor was submerged again in a second Hg dissolution of 20 ppb. As can be seen in Figure 9, the comparison of two measurements of the same optical fiber sensor against two different samples of mercury (20 ppb) is represented. The first measurement corresponds to the freshly fabricated sensor that was deposited in a first mercury sample (20 ppb). After reacting to mercury was deposited in the regenerative solution that allowed the sensor to recover the initial conditions, so it was deposited in a new sample of mercury (20 ppb), thus obtaining the second measurement, yielding a very similar wavelength shift as in the first measurement.

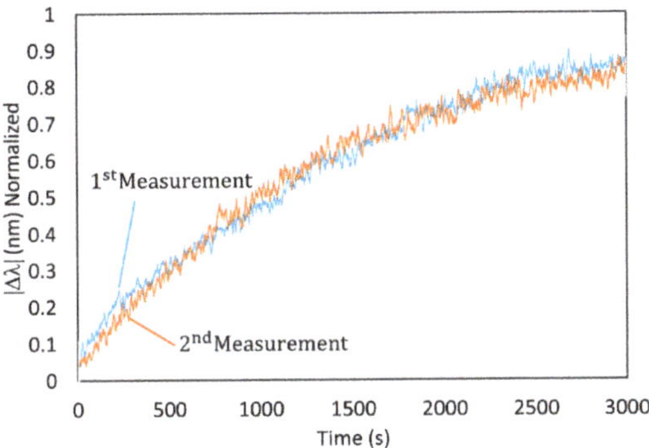

Figure 9. Regeneration of the optical fiber sensor. Repeatability of the sensor under 20 ppb of mercury after its regeneration in a dilute nitric acid solution. The final wavelength-shift is very similar in both cases, and the response slope is even more stable.

3.5. Cross Sensitivity

Finally, the selectivity of optical fiber sensor against different heavy metal ions (Fe^{2+}, Ni^{2+}, Pb^{2+}, Cd^{2+}, and Zn^{2+}) is also studied the same sensor has been exposed to the same concentration (20 ppb) of the different metal ions, and all solutions were prepared using the same protocol as in the previous mercury tests (PB + DMAB @pH 7.6). The results of the final wavelength shift after the immersion in the different ion solutions are showed in Figure 10.

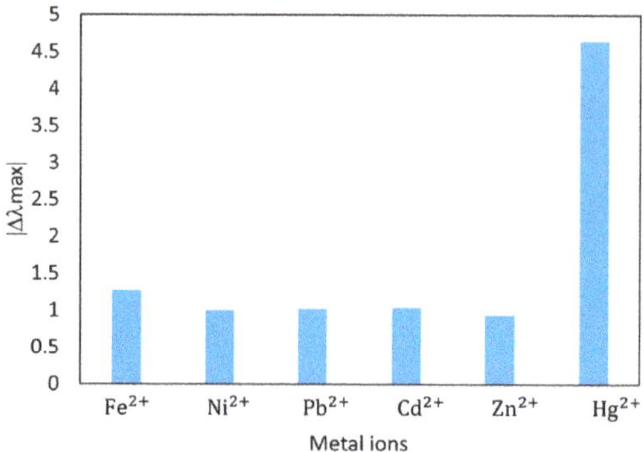

Figure 10. Selectivity analysis against the most common heavy ions. The measurements were carried out using the same sampling preparation process, and with the same concentration (20 ppb).

As it can be seen in Figure 10, the proposed optical fiber sensor showed a significantly higher response to the presence of mercury compared to the other metal ions, nearly 400% higher than the second more reactive cross-contaminant (in this case iron) enjoying a high selectivity towards mercury.

The metal ions analyzed have detection limits allowed in water higher than of mercury [28], for example Ni (15 ppb), Fe (6.2 ppb) among others. Therefore, in normal samples, our sensor would be more selective to mercury than to other metal ions. A further study would require the evaluation of the present devices in real aquifer water samples, but this is out of the scope of the present contribution. This work presents a competitive approach for a mercury optical fiber sensor, with a simple and direct measurement of mercury in water.

4. Conclusions

In this work, a simple and highly sensitive mercury optical fiber sensor has been proposed. Its sensing mechanism is based on the wavelength shift of the AuNPs LSPR, thanks to the strong chemical affinity of mercury towards gold. The gold nanoparticles were obtained by synthesis method of the PAA-capped AuNPs, using PAA as a stabilizing agent and DMAB as reducing agent. The LBL nano-assembly technique was used for the incorporation of gold nanoparticles onto optical fiber in a stable sensitive thin-film, $(PAH/PAA+AuNPs)_6$. The absolute wavelength-shift was a reliable and robust signal with a relatively long response time of around one hour. In order to obtain faster measurements, the slope of the wavelength variation proved to give reliable results in only 8 min. It is possible to reuse the sensor, something that reduces costs and manufacturing time. In addition, this sensor showed low cross sensitivity towards other metal ions. It was achieved a limit of detection of 0.7 ppb, which is lower than the standard limits recommended by the European Union (1 ppb) and US-EPA (2 ppb). The sensor proposed in this work could be competitive alternative for mercury detection, a problem of global concern.

Author Contributions: Conceptualization and Methodology, M.E.M.-H., J.G. and F.J.A.; Investigation and Validation, M.E.M.-H.; Writing-Original Draft Preparation, M.E.M.-H. and J.G.; Writing-Review & Editing, F.J.A.; Supervision, J.G. and F.J.A.; Project Administration and Funding Acquisition, F.J.A.

Funding: This work was supported in part by the Spanish Agencia Estatal de Investigación (AEI) and the European Regional Development Fund (ERDF) under the Project TEC2016-79367-C2-2-R, and Public University of Navarra pre-doctoral research grants.

Conflicts of Interest: The authors declare no conflict of interest.

References

1. Zou, Y.; Zhang, Y.; Xie, Z.; Luo, S.; Zeng, Y.; Chen, Q.; Liu, G.; Tian, Z. Improved sensitivity and reproducibility in electrochemical detection of trace mercury (II) by bromide ion & electrochemical oxidation. *Talanta* **2019**, *203*, 186–193.
2. Gong, J.; Zhou, T.; Song, D.; Zhang, L. Monodispersed Au nanoparticles decorated graphene as an enhanced sensing platform for ultrasensitive stripping voltammetric detection of mercury (II). *Sens. Actuators B Chem.* **2010**, *150*, 491–497. [CrossRef]
3. Deshmukh, M.A.; Celiesiute, R.; Ramanaviciene, A.; Shirsat, M.D.; Ramanavicius, A. EDTA_PANI/SWCNTs nanocomposite modified electrode for electrochemical determination of copper (II), lead (II) and mercury (II) ions. *Electrochim. Acta* **2018**, *259*, 930–938. [CrossRef]
4. Bansod, B.K.; Kumar, T.; Thakur, R.; Rana, S.; Singh, I. A review on various electrochemical techniques for heavy metal ions detection with different sensing platforms. *Biosens. Bioelectron.* **2017**, *94*, 443–455. [CrossRef] [PubMed]
5. Tan, S.Y.; Lee, S.C.; Okazaki, T.; Kuramit, H.; Abd-Rahman, F. Detection of mercury (II) ions in water by polyelectrolyte–gold nanoparticles coated long period fiber grating sensor. *Opt. Commun.* **2018**, *419*, 18–24. [CrossRef]
6. Priyadarshini, E.; Pradhan, N. Gold nanoparticles as efficient sensors in colorimetric detection of toxic metal ions: A review. *Sens. Actuators B Chem.* **2017**, *238*, 888–902. [CrossRef]
7. Srivastava, S.K.; Arora, V.; Sapra, S.; Gupta, B.D. Localized Surface Plasmon Resonance-Based Fiber Optic U-Shaped Biosensor for the Detection of Blood Glucose. *Plasmonics* **2012**, *7*, 261–268. [CrossRef]
8. Rycenga, M.; Cobley, C.M.; Zeng, J.; Li, W.; Moran, C.H.; Zhang, Q.; Qin, D.; Xia, Y. Controlling the synthesis and assembly of silver nanostructures for plasmonic applications. *Chem. Rev.* **2011**, *111*, 3669–3712. [CrossRef] [PubMed]
9. Rivero, P.J.; Hernaez, M.; Goicoechea, J.; Matías, I.R.; Arregui, F.J. A comparative study in the sensitivity of optical fiber refractometers based on the incorporation of gold nanoparticles into layer-by-layer films. *Int. J. Smart Sens. Intell. Syst.* **2015**, *8*, 822–841. [CrossRef]
10. Jia, S.; Bian, C.; Sun, J.; Tong, J.; Xia, S. A wavelength-modulated localized surface plasmon resonance (LSPR) optical fiber sensor for sensitive detection of mercury(II) ion by gold nanoparticles-DNA conjugates. *Biosens. Bioelectron.* **2018**, *114*, 15–21. [CrossRef]
11. Haynes, C.L.; van Duyne, R.P. Nanosphere lithography: A versatile nanofabrication tool for studies of size-dependent nanoparticle optics. *J. Phys. Chem. B* **2001**, *105*, 5599–5611. [CrossRef]
12. Shukla, G.M.; Punjabi, N.; Kundu, T.; Mukherji, S. Optimization of Plasmonic U-shaped Optical Fiber Sensor for Mercury Ions Detection using Glucose Capped Silver Nanoparticles. *IEEE Sens. J.* **2019**, *19*, 3224–3231. [CrossRef]
13. Sadani, K.; Nag, P.; Mukherji, S. LSPR based optical fiber sensor with chitosan capped gold nanoparticles on BSA for trace detection of Hg (II) in water, soil and food samples. *Biosens. Bioelectron.* **2019**, *134*, 90–96. [CrossRef] [PubMed]
14. Sai, V.V.R.; Kundu, T.; Mukherji, S. Novel U-bent fiber optic probe for localized surface plasmon resonance based biosensor. *Biosens. Bioelectron.* **2009**, *24*, 2804–2809. [CrossRef]
15. Abu-Ali, H.; Nabok, A.; Smith, T.J. Development of novel and highly specific ssDNA-aptamer-based electrochemical biosensor for rapid detection of mercury (II) and lead (II) ions in water. *Chemosensors* **2019**, *7*, 27. [CrossRef]
16. Fayazi, M.; Taher, M.A.; Afzali, D.; Mostafavi, A. Fe3O4 and MnO2 assembled on halloysite nanotubes: A highly efficient solid-phase extractant for electrochemical detection of mercury(II) ions. *Sens. Actuators B Chem.* **2016**, *228*, 1–9. [CrossRef]
17. Paulauskas, A.; Selskis, A.; Bukauskas, V.; Vaicikauskas, V.; Ramanavicius, A.; Balevicius, Z. Real time study of amalgam formation and mercury adsorption on thin gold film by total internal reflection ellipsometry. *Appl. Surf. Sci.* **2018**, *427*, 298–303. [CrossRef]
18. Vasjari, M.; Shirshov, Y.M.; Samoylov, A.V.; Mirsky, V.M. SPR investigation of mercury reduction and oxidation on thin gold electrodes. *J. Electroanal. Chem.* **2007**, *605*, 73–76. [CrossRef]

19. Sabri, N.; Aljunid, S.A.; Salim, M.S.; Fouad, S. Fiber Optic Sensors: Short Review and Applications. In *Recent Trends in Physics of Material Science and Technology*; Gaol, F.L., Shrivastava, K., Akhtar, J., Eds.; Springer: Singapore, 2015; pp. 299–311.
20. Elosua, C.; Arregui, F.J.; del Villar, I.; Ruiz-Zamarreño, C.; Corres, J.M.; Bariain, C.; Goicoechea, J.; Hernaez, M.; Rivero, P.J.; Socorro, A.B.; et al. Micro and nanostructured materials for the development of optical fibre sensors. *Sensors* **2017**, *17*, 2312. [CrossRef]
21. Jia, S.; Bian, C.; Tong, J.H.; Sun, J.Z.; Xia, S.H. A Fiber-optic Sensor Based on Plasmon Coupling Effects in Gold Nanoparticles Core-satellites Nanostructure for Determination of Mercury Ions (II). *Chin. J. Anal. Chem.* **2017**, *45*, 785–790. [CrossRef]
22. Schopf, C.; Martín, A.; Schmidt, M.; Iacopino, D. Investigation of Au-Hg amalgam formation on substrate-immobilized individual Au nanorods. *J. Mater. Chem. C* **2015**, *3*, 8865–8872. [CrossRef]
23. Goicoechea, J.; Rivero, P.J.; Sada, S.; Arregui, F.J. Self-Referenced Optical Fiber Sensor for Hydrogen Peroxide Detection based on LSPR of Metallic Nanoparticles in Layer-by-Layer Films. *Sensors* **2019**, *19*, 3872. [CrossRef] [PubMed]
24. Decher, G.; Eckle, M.; Schmitt, J.; Struth, B. Layer-by-layer assembled multicomposite films. *Curr. Opin. Colloid Interface Sci.* **1998**, *3*, 32–39. [CrossRef]
25. Fuzzy, D.G. Nanoassemblies: Toward layered polymeric multicomposites. *Science* **1997**, *277*, 1232–1237.
26. Baba, Y.; Ohe, K.; Kawasaki, Y.; Kolev, S.D. Adsorption of mercury(II) from hydrochloric acid solutions on glycidylmethacrylate-divinylbenzene microspheres containing amino groups. *React. Funct. Polym.* **2006**, *66*, 1158–1164. [CrossRef]
27. Jenkins, D.W. *Biological Monitoring of Toxic Trace Metals*; EPA Report 600/S3-80-090; United States Environmental Protection Agency: Washington, DC, USA, 1980; pp. 1–9.
28. Wang, T.; Kang, D.H.; Yu, Y.J.; Gu, J.H.; Yang, L.R. Determination of macro and trace elements in rare earth magnesium cast iron by inductively coupled plasma atomic emission spectrometry. *Yejin Fenxi/Metall. Anal.* **2012**, *32*, 66–69.
29. USEPA. *National Recommended Water Quality Criteria*; 4304T; United States Environmental Protection Agency: Washington, DC, USA, 2009; Volume 1.
30. European Commission. Council Directive 98/83/EC of 3 November 1998 on the Quality of Water Intended for Human Consumption. *Off. J. Eur. Communities* **1998**, *41*, 34–54.

Communication

Analysis of Trace Metals in Human Hair by Laser-Induced Breakdown Spectroscopy with a Compact Microchip Laser

Makoto Nakagawa [1] and Yuji Matsuura [1,2,*]

[1] Graduate School of Engineering, Tohoku University, 6-6-05 Aoba, Sendai 980-8579, Japan; makoto.nakagawa.p1@dc.tohoku.ac.jp
[2] Graduate School of Biomedical Engineering, Tohoku University, 6-6-05 Aoba, Sendai 980-8579, Japan
* Correspondence: yuji@ecei.tohoku.ac.jp; Tel.: +81-22-795-7108

Abstract: A laser-induced breakdown spectroscopy (LIBS) system using a microchip laser for plasma generation is proposed for in-situ analysis of trace minerals in human hair. The LIBS system is more compact and less expensive than conventional LIBS systems, which use flashlamp-excited Q-switched Nd:YAG lasers. Focusing optics were optimized using a Galilean beam expander to compensate for the low emitted pulse energy of the microchip laser. Additionally, hundreds of generated LIBS spectra were accumulated to improve the signal-to-noise ratio of the measurement system, and argon gas was injected at the irradiation point to enhance plasma intensity. LIBS spectra of human hair in the UV to near IR regions were investigated. Relative mass concentrations of Ca, Mg, and Zn were analyzed in hairs obtained from five subjects using the intensity of C as a reference. The results coincide well with those measured via inductively coupled argon plasma mass spectrometry. The lowest detectable concentrations of the measured LIBS spectra were 9.0 ppm for Mg, 27 ppm for Zn, and 710 ppm for Ca. From these results, we find that the proposed LIBS system based on a microchip laser is feasible for the analysis of trace minerals in human hair.

Keywords: laser-induced plasma spectroscopy; microchip laser; hair analysis

Citation: Nakagawa, M.; Matsuura, Y. Analysis of Trace Metals in Human Hair by Laser-Induced Breakdown Spectroscopy with a Compact Microchip Laser. *Sensors* **2021**, *21*, 3752. https://doi.org/10.3390/s21113752

Academic Editor: Gabriela Kuncová

Received: 13 April 2021
Accepted: 25 May 2021
Published: 28 May 2021

1. Introduction

Laser-induced breakdown spectroscopy (LIBS) is a technique that measures emission spectra from luminous plasma generated by irradiation with nano-, pico-, and femto-second laser pulses and is useful for multi-elemental analysis of various target materials [1,2]. A microchip laser [3–6] that emits a pulse energy of hundreds of microjoules has become popular because it makes the LIBS system more compact and lower in cost than conventional systems, which use flashlamp-excited Q-switched Nd:YAG lasers [7,8]. Portable LIBS systems with microchip lasers and compact fiber-coupled spectrometers have been developed [9]. These compact systems have been used for quantitative analysis of steel composition [10,11] and aluminum alloys [12,13].

LIBS techniques are useful for qualitative and quantitative analysis of biological samples and have been applied in the diagnosis of some diseases, such as cancer [14]. In biomedical applications, one of the advantages of the LIBS technique is that pretreatment of samples is not required, unlike in other elemental analysis methods such as inductively coupled argon plasma-atomic emission spectroscopy (ICP-AES) or mass spectrometry (ICP-MS) [15]. For healthcare applications, such as nutritional status monitoring, analysis of easily harvested biological specimens, such as nails and hair, is useful. ICP-AES and ICP-MS have already been applied to the analysis of a variety of biological samples, including nails and hair [16]. However, as mentioned above, ICP-AES and ICP-MS need relatively complicated pretreatment processes and, therefore, real-time analysis is difficult. Additionally, the large-scale and high-cost equipment that is necessary for those analysis techniques is not cost-effective for most healthcare applications. Therefore, many groups have proposed LIBS techniques for the analysis of biological samples. LIBS spectra of fingernails have been measured for the diagnosis of diseases, such as diabetes. As nails are relatively hard tissues with high mechanical and chemical strength, stable LIBS measurements are possible [17–20].

Hair is another target tissue for LIBS measurements because the concentrations of trace elements in hair are generally higher than that in other biological tissues. Haruna detected Ca in human hair using LIBS and qualitatively analyzed Ca variation with age and sex [21]. Corsi et al. quantitatively analyzed mineral content (Mg, K, Ca, Na, and Al) in human hair and compared their results with those obtained through a commercial analytical laboratory [22]. Emara measured trace elements in horsehair and compared the results with those obtained via atomic absorption spectroscopy [23]. More recently, Zhang combined LIBS with ultrasound-assisted alkali dissolution to analyze Zn and Cu more accurately in human hair [24]. In these LIBS applications, conventional Q-switched Nd: YAG lasers are used. This is because conventional Q-switched Nd: YAG lasers can obtain relatively high plasma intensity by irradiation with pulse energy of more than 10 mJ, which is necessary to detect trace elements in hair. The use of these lasers makes the LIBS system large in scale and not easy to handle. In addition, irradiation with high energy pulses easily induces severe damage to hair and sometimes the hair is torn off, which can cause changes in the detected signal while obtaining the LIBS spectra.

In this paper, we propose LIBS analysis of human hair while using a microchip laser for plasma generation. To obtain plasma intensity that is sufficiently high for trace element analysis, we optimally designed the focusing optics. We then accumulate hundreds of LIBS spectra generated by the microchip laser emitting at high repetition rate pulses while Argon gas is injected at the irradiation point to enhance plasma intensity. Table 1 presents the typical concentrations of trace elements contained in the black hair of Japanese adults [25]. Among these elements, we analyzed the relative mass concentrations of Ca, Mg, and Zn of hairs obtained from five subjects while using the intensity of C as a reference. The concentration results coincide well with those measured via ICP-MS.

Table 1. Typical concentrations of trace elements contained in the black hair of Japanese adults [25].

Element	Ca	Zn	Mg	Fe	Al	K	Na
Concentration (ppm)	810	179	164	144	130	70	65

2. Materials and Methods

In our experiment, a passively Q-switched microchip laser (L11038–11, Hamamatsu Photonics, Hamamatsu, Japan) emitting optical pulses with an energy of 2 mJ and pulse duration of 1 ns at a wavelength of 1064 nm was used as the light source. The measurement setup is shown in Figure 1. The experiment was conducted under atmospheric pressure. The plasma generated from the sample was delivered to the spectrometer by a step-index silica-glass fiber with a core diameter of 600 µm and NA of 0.22. The distal end of the fiber was located at a distance of 2 mm from the irradiation point. The spectra were measured using a fiber-coupled spectrometer (HR2000+, Ocean Insight, Orlando, FL, USA) which was synchronized with the laser pulses using an external pulse generator. The integration time of the spectrometer was set to 1 ms, which was the shortest possible setting, and was not gated. Therefore, the plasma emission generated by a nanosecond optical pulse was completely detected. Three types of spectrometers were used to obtain LIBS spectra from the ultraviolet (UV) to visible (Vis) regions. With the combined use of three spectrometers, the wavelength ranges of 200–343 nm, 355–474 nm, and 480–597 nm were covered with a wavelength resolution of 0.14 nm.

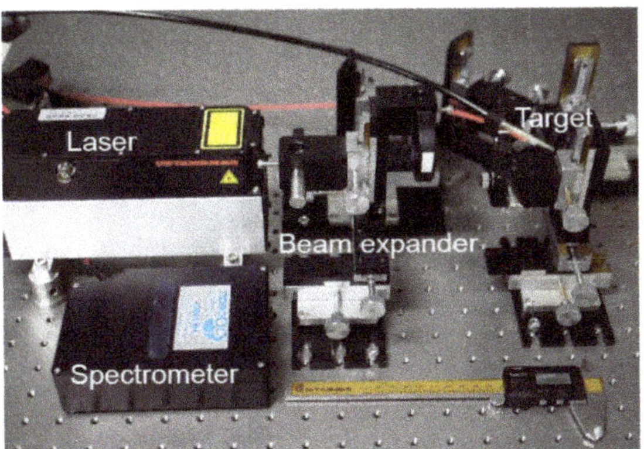

Figure 1. Measurement setup.

Dozens of human hairs were bundled and fixed on a metal substrate to create samples. The substrate had a hole with a diameter of 12 mm at the center to avoid generating the plasma from the underlying substrate. The hairs were washed twice with pure water after degreasing with acetone. When irradiating with optical pulses, the sample was moved slowly along the longitudinal direction at a speed of approximately 2 cm/min to avoid severely damaging the hair sample. Our protocol was approved by the ethical committee on the Use of Humans as Experimental Subjects of Tohoku University (No. 20A-29), and informed consent was obtained from the examinees.

We first focused the laser beam, using a spherical BK7 lens with a focal length of 100 mm, to obtain the LIBS spectrum shown in Figure 2. For each measurement, 300 consecutive spectra were accumulated. Because of the limited data acquisition speed of the spectrometer, the repetition rate of the laser was set to 10 Hz, and thus, it took 30 s for a single measurement. To identify elements from the LIBS spectra, we referred to the NIST Database [26] and OSCAR Database [27]. Although we confirmed emission peaks of Ca at Ca I: 422.7 and 445.5 nm, Ca II: 393.3 and 396.8 nm, and of the C–N bond at 388.29 nm, we could not detect peaks of another material because of the low signal-to-noise ratio (SNR) of the LIBS system.

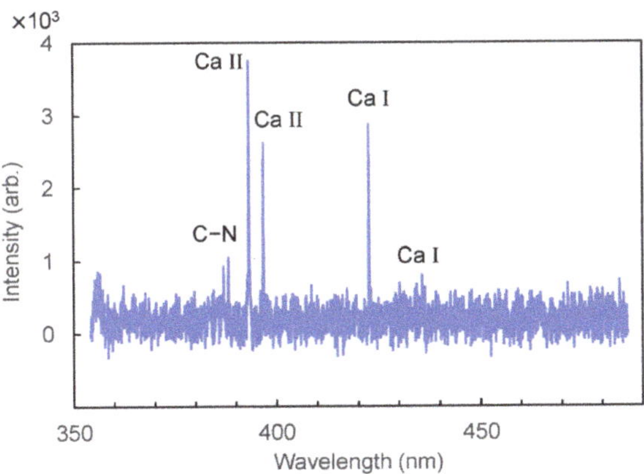

Figure 2. Laser-induced breakdown spectroscopy (LIBS) spectrum of hair measured by a focusing system with a single lens.

To improve the SNR, we designed focusing optics to increase the optical power density at the focusing spot. To obtain a more focused beam spot, we built a Galilean beam expander with 5× magnification. The focused beam size was reduced to 0.12 mm from 0.37 mm by introducing the expander. The power density at the focal spot was approximately 4.5×10^9 W/cm^2. Figure 3 shows a LIBS spectrum of hair measured with and without the beam expander. We found that the peak intensity was enhanced by approximately 10 times.

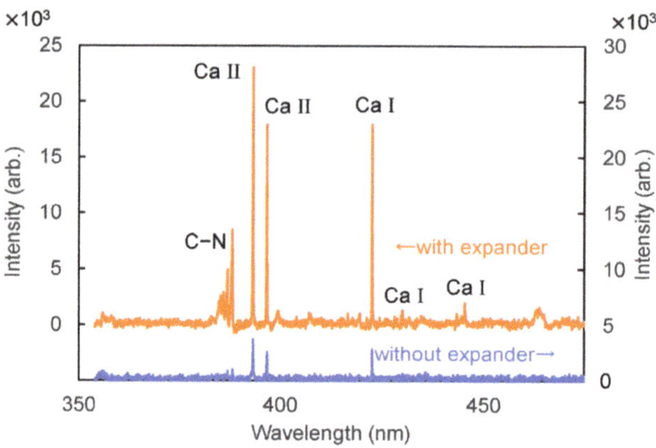

Figure 3. LIBS spectra of hair measured with and without an expander.

To further increase the sensitivity, we injected argon gas onto the focusing spot. It was reported that the optical emission intensity of plasma induced by laser irradiation is enhanced in the argon atmosphere [28]. One of the reasons for this is that the energy decay of free electrons in plasma is suppressed in an argon atmosphere. In addition, high plasma temperature is maintained in argon because of its lower thermal conductivity. Figure 4 shows LIBS spectra of human hair measured with and without argon injection. Argon gas was injected from a nozzle set at a distance of 2 mm from the focused spot with a flow rate of 1 L/min. It was found that the peak intensity was enhanced approximately 1.8 times using argon injection; small peaks that can be attributed to C–N bond were observed at around 358 nm.

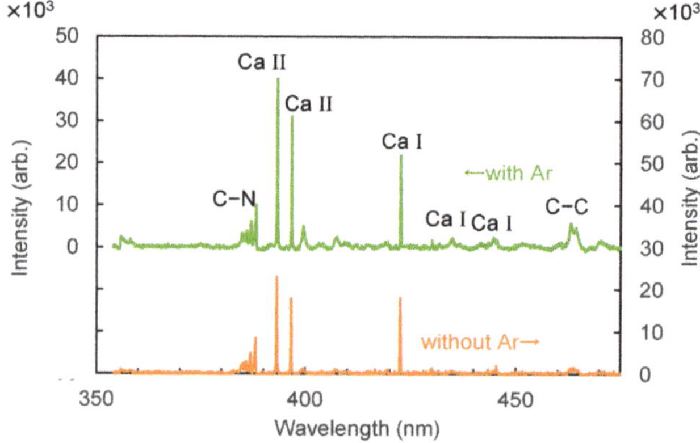

Figure 4. LIBS spectra of hair measured with and without argon gas injection.

To find the optimum number of spectrum accumulation in a LIBS measurement, we changed the accumulation number from 1 to 900 and measured the LIBS spectra of human hair. Figure 5 shows the correlation between the SNR of the obtained spectra and the accumulation number. The SNR was defined as (Peak area of Ca II at 393.3 nm)/[(standard deviation of background signal) × (full width of the Ca II peak at half maximum)]. We found that the SNR was almost saturated at an accumulation number of 300, and therefore, we set the accumulation number to 300 in consideration of the measurement time.

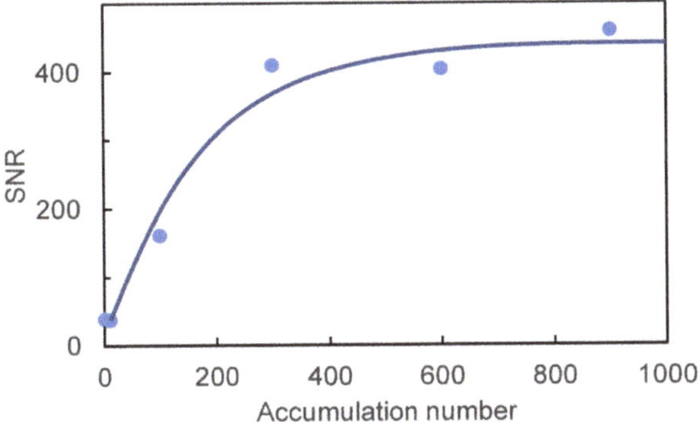

Figure 5. Correlation between signal-to-noise ratio (SNR) of obtained spectra and accumulation number.

3. Results and Discussions

Figure 6 shows a LIBS spectrum of human hair measured from 480 to 597 nm. In this wavelength region, emission peaks for the C–C bond at 512.8 and 516.4 nm, Ba I at 553.5 nm, and Na I at 589.6 nm are observed. Figure 7 shows a spectrum measured at 200–340 nm. In the measurement results for the UV region, we found that there was less background noise from calcium emission; thus, we changed the focal length of the lens from 100 mm to 50 mm to further increase the energy density at the focused spot. The focal spot size was reduced to 0.09 mm from 0.12 mm. In this region, we observed peaks of C I (247.8 nm), Mg II (279.5 nm), and Ca II (315.8 nm). Figure 8 shows an enlarged spectrum at around 330 nm. Although the peak intensities are relatively low, we found that the peaks coincide with that of Zn I at 328.26, 330.26, and 334.51 nm. Since it was found that the peak at 330.26 nm was affected by the peak of Na I at 330.3 nm, we hereafter use the peak at 328.26 nm for analysis of Zn I, which is one of the trace minerals essential for human life.

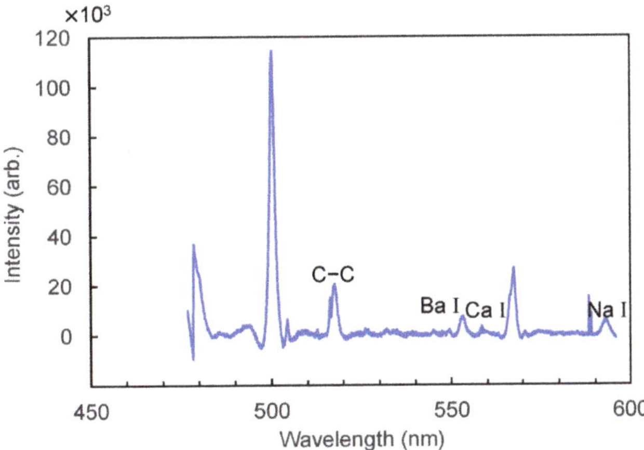

Figure 6. LIBS spectrum of hair measured in the visible wavelength region.

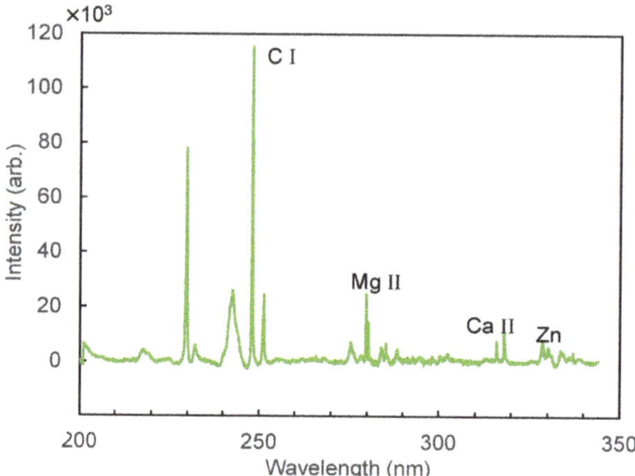

Figure 7. LIBS spectrum of hair measured in the ultraviolet wavelength region.

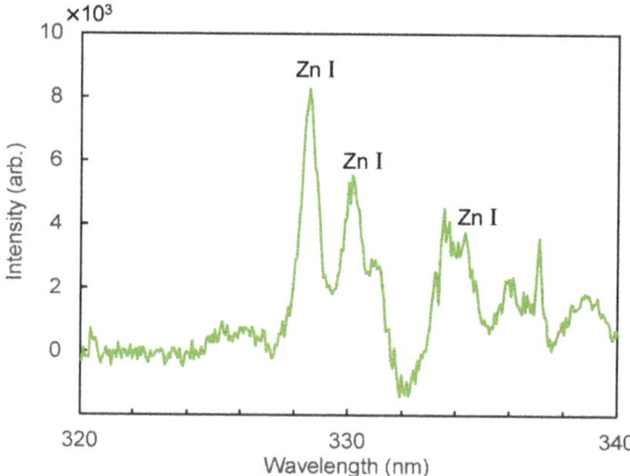

Figure 8. An enlarged LIBS spectrum of hair measured at around 330 nm.

We performed LIBS measurements to observe individual differences in the relative concentrations of the trace minerals. We collected hair samples from five volunteers, aged 23–25, and compared the results of LIBS analysis to those obtained via ICP-MS analysis. LIBS measurements were used to observe individual differences in the relative concentrations of these trace minerals. As the concentration of trace elements in hair depends on the position along the hair length, we analyzed the hair around 2.5 cm from the root for all the measurements. We utilized the commercial service of the Kyorin Preventive Medicine Institute [29] for ICP-MS analysis of hair samples that were taken from the same subjects at the same time. For ICP-MS analysis, the same part of hair as described above was used for comparisons.

As the intensity of observed peaks varies widely because of the small diameter of the hair samples, the relative concentrations obtained via LIBS analysis were calculated by setting the peak intensity of C I as the reference because C is a primary composition component of hair; as such, the individual difference should be relatively small. Table 2 shows the observed peak intensities of Mg II (279.5 nm), Zn I (328.26), and Ca II (315.8 nm) measured for samples taken from the five subjects. The coefficients of variation (CV) were calculated from the results of three measurements for each subject, and we confirmed that the variations reduced considerably upon using the peak intensity of C as a reference. We also tried to normalize the peak intensities by the total emission intensity of measured LIBS spectra. However, we did not obtain better results because of the relatively large background noises in the measurement. The calculated total emission intensity largely changed with baseline correction processing.

Table 2. Effect of normalization based on the peak intensity of C I at 247.8 nm. In the table, "X" and "X/C" show the coefficients of variations (CV) before and after normalization, respectively.

Subject		1			2			#3			4			5		
Element X		Mg	Zn	Ca	Mg	Zn	Ca	Mg	Zn	Ca	Mg	Zn	Ca	Mg	Zn	Ca
CV	X	0.126	0.350	0.134	0.159	0386	0.209	0.249	0.573	0.285	0.034	0.053	0.141	0.107	0.151	0.157
	X/C	0.043	0.189	0.037	0.102	0.212	0.046	0.152	0.505	0.055	0.044	0.022	0.082	0.023	0.170	0.073

Figure 9 shows the relative mass concentrations of Mg, Zn, and Ca measured for the five subjects compared with the absolute concentrations analyzed via ICP-MS. In the LIBS results in Figure 9, the dots are the average values of three measurements, and the error bars show the minimum and the maximum measured values. The measurement variability was sufficiently small to see individual differences and the trends between the subjects coincided with the results of ICP-MS analysis.

We confirmed good linearity between the results of the LIBS and ICP-MS methods, and the determination coefficient R^2 was 0.921 for Mg, 0.670 for Zn, and 0.952 for Ca. We observed relatively small correlation for Zn; this may be because of the small peak intensity compared to the ones of Mg and Ca. The lowest detectable concentrations, defined by SNR = 3, were 9.0 ppm for Mg, 27 ppm for Zn, and 710 ppm for Ca. Since these values are lower than typical concentrations of these trace minerals, we confirmed the feasibility of the proposed LIBS using a microchip laser for analysis of relative mass concentrations in human hair. In the above analysis, we did not consider the variability of plasma temperature, electron density, and upper energy levels of the observed transitions [30]. For more accurate analysis of trace elements in human hair, corrections based on these factors may be necessary.

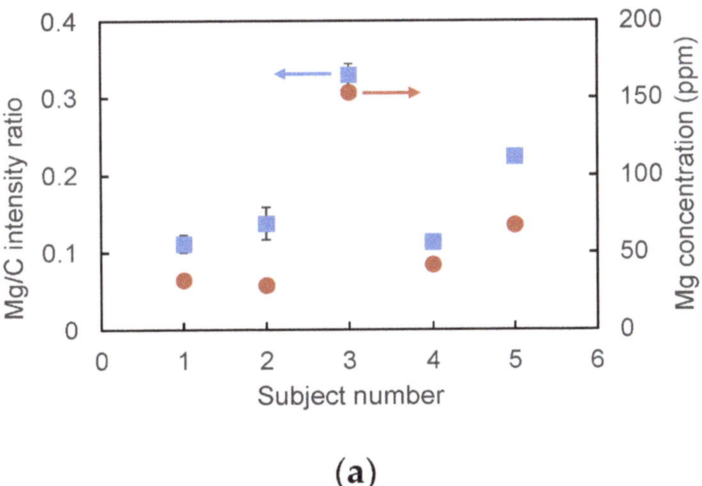

Figure 9. *Cont.*

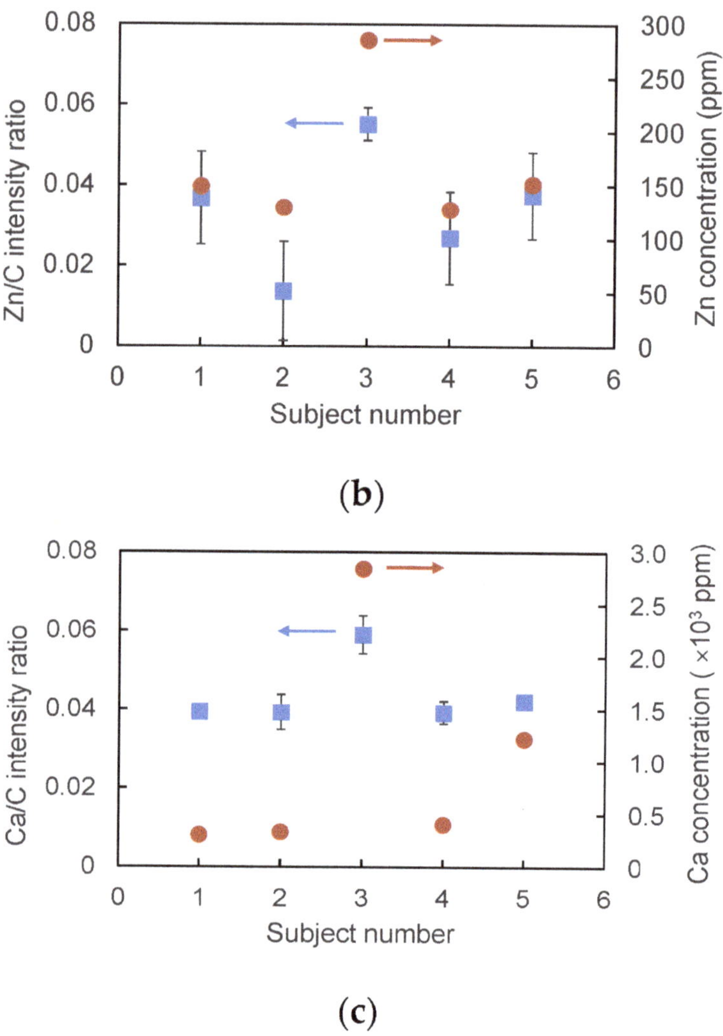

Figure 9. Relative mass concentrations of (**a**) Mg, (**b**) Zn, and (**c**) Ca measured for the five subjects and compared with the absolute concentrations analyzed via ICP-MS.

4. Conclusions

As a compact and low-cost LIBS system for in-situ analysis of trace minerals in human hair, we proposed a system using a microchip laser for plasma generation. Since the pulse energy emitted from a microchip laser is lower than that of conventional flashlamp-excited Q-switched Nd:YAG lasers, we optimally designed the focusing optics utilizing a Galilean beam expander to obtain plasma intensity sufficiently high for analysis of trace elements in hair. Additionally, we accumulated hundreds of LIBS spectra to improve the SNR of the measurement system and injected argon gas on the irradiation point to enhance the plasma intensity. After investigating LIBS spectra of human hair in the UV to near IR regions, we focused on the spectra in the UV region because of the location of emission peaks of Mg and Zn, which are trace minerals essential for human life.

We analyzed relative mass concentrations of Ca, Mg, and Zn in hairs obtained from five subjects while using the intensity of the C peak as a reference. The results coincided well with those measured via ICP-MS. We estimated the lowest detectable concentrations from the SNR of the measured LIBS spectra: 9.0 ppm for Mg, 27 ppm for Zn, and 710 ppm for Ca. From these results, we have concluded that the proposed LIBS system based on a microchip laser is feasible for the analysis of trace minerals in human hair.

Owing to the low-cost and compact proposed system, we expect that it will be a useful biomedical sensor for health-care applications based on non-invasive and real time analysis of hair.

Author Contributions: M.N. performed the experiments and analyzed the data; Y.M. conceived and designed the experiments. All authors have read and agreed to the published version of the manuscript.

Funding: This research was funded by JSPS KAKENHI Grant Numbers JP 20H00231 and JP 20K12615.

Institutional Review Board Statement: The study was conducted according to the guidelines of the Declaration of Helsinki, and approved by the he Ethical Committee on the Use of Humans as Experimental Subjects of Tohoku University (20A-29, 9 October 2020).

Informed Consent Statement: Informed consent was obtained from all subjects involved in the study.

Data Availability Statement: The data presented in this study are available on request from the corresponding author. The data are not publicly available due to privacy restrictions.

Conflicts of Interest: The authors declare no conflict of interest.

References

1. Cremers, D.A.; Radziemski, L.J. *Handbook of Laser-Induced Breakdown Spectroscopy*; John Wiley & Sons: Wes Sussex, UK, 2006. [CrossRef]
2. Musazzi, S.; Perini, U. *Laser-Induced Breakdown Spectroscopy: Theory and Applications*; Springer: Berlin/Heidelberg, Germany, 2014. [CrossRef]
3. Zayhowski, J.J. Passively Q-switched Nd: YAG microchip lasers and applications. *J. Alloy. Compd.* **2000**, *46*, 393–400. [CrossRef]
4. Sakai, H.; Kan, H.; Taira, T. 1 MW peak power single-mode high-brightness passively Q-switched Nd^{3+}: YAG microchip laser. *Opt. Express* **2008**, *16*, 19891–19899. [CrossRef]
5. Merten, J.A.; Ewusi-Annan, E.; Smith, B.W.; Omenetto, N. Optimizing gated detection in high-jitter kilohertz powerchip laser-induced breakdown spectroscopy. *J. Anal. At. Spectrom.* **2014**, *29*, 571–577. [CrossRef]
6. Merten, J.A.; Smith, B.W.; Omenetto, N. Local thermodynamic equilibrium considerations in powerchip laser-induced plasmas. *Spectrochim. Acta Part B At. Spectrosc.* **2013**, *83–84*, 50–55. [CrossRef]
7. Singh, J.P.; Thakur, S.N. *Laser-Induced Breakdown Spectroscopy*; Elsevier: Amsterdam, The Netherlands, 2007; pp. 173–196. [CrossRef]
8. Amponsah-Manager, K.; Omenetto, N.; Smith, B.W.; Gornushkin, I.B.; Winefordner, J.D. Microchip laser ablation of metals: Investigation of the ablation process in view of its application to laser-induced breakdown spectroscopy. *J. Anal. At. Spectrom.* **2005**, *20*, 544–551. [CrossRef]
9. Gornushkin, I.B.; Amponsah-Manager, K.; Smith, B.W.; Omenetto, N.; Winefordner, J.D. Microchip laser-induced breakdown spectroscopy: A preliminary feasibility investigation. *Appl. Spectrosc.* **2004**, *58*, 762–769. [CrossRef]
10. Lopez-Moreno, C.; Amponsah-Manager, K.; Smith, B.W.; Gornushkin, I.B.; Omenetto, N.; Palanco, S.; Lasernab, J.J.; Winefordner, J.D. Quantitative analysis of low-alloy steel by microchip laser induced breakdown spectroscopy. *J. Anal. At. Spectrom.* **2005**, *20*, 552–556. [CrossRef]
11. Gonzaga, F.B.; Pasquini, C. A compact and low cost laser induced breakdown spectroscopic system: Application for simultaneous determination of chromium and nickel in steel using multivariate calibration. *Spectrochim. Acta Pt. B At. Spec.* **2012**, *69*, 20–24. [CrossRef]
12. Freedman, A.; Iannarilli, F.J.; Wormhoudt, J.C. Aluminum alloy analysis using microchip-laser induced breakdown spectroscopy. *Spectrochim. Acta Pt. B At. Spec.* **2005**, *60*, 1076–1082. [CrossRef]
13. Cristoforetti, G.; Legnaioli, S.; Palleschi, V.; Salvetti, A.; Tognoni, E.; Benedetti, P.A.; Brioschi, F.; Ferrario, F. Quantitative analysis of aluminium alloys by low-energy, high-repetition rate laser-induced breakdown spectroscopy. *J. Anal. At. Spectrom.* **2006**, *21*, 697–702. [CrossRef]
14. Gaudiuso, R.; Melikechi, N.; Abdel-Salam, Z.A.; Harith, M.A.; Palleschi, V.; Motto-Ros, V.; Busser, B. Laser-induced breakdown spectroscopy for human and animal health: A review. *Spectrochim. Acta Pt. B At. Spec.* **2019**, *152*, 123–148. [CrossRef]
15. Thompson, M.; Walsh, J.N. *Handbook of Inductively Coupled Plasma Spectrometry*, 2nd ed.; Springer: Berlin/Heidelberg, Germany, 1989. [CrossRef]
16. Vanhoe, H. A review of the capabilities of ICP-MS for trace element analysis in body fluids and tissues. *J. Trace Elem. Electrol. Health Dis.* **1993**, *7*, 131–139. Available online: http://europepmc.org/abstract/MED/8155984 (accessed on 9 December 2020).

17. Bahreini, M.; Hosseinimakarem, Z.; Tavassoli, S.H. A study of association between fingernail elements and osteoporosis by laser-induced breakdown spectroscopy. *J. Appl. Phys.* **2012**, *112*, 054701. [CrossRef]
18. Hosseinimakarem, Z.; Tavassoli, S.H. Analysis of human nails by laser-induced breakdown spectroscopy. *J. Biomed. Opt.* **2011**, *16*, 057002. [CrossRef]
19. Hamzaoui, S.; Khleifia, R.; Jaïdane, N.; Lakhdar, Z.B. Quantitative analysis of pathological nails using laser-induced breakdown spectroscopy (LIBS) technique. *Lasers Med. Sci.* **2011**, *26*, 79–83. [CrossRef]
20. Bahreini, M.; Ashrafkhani, B.; Tavassoli, S.H. Discrimination of patients with diabetes mellitus and healthy subjects based on laser-induced breakdown spectroscopy of their fingernails. *J. Biomed. Opt.* **2013**, *18*, 107006. [CrossRef]
21. Haruna, M.; Ohmi, M.; Nakamura, M.; Morimoto, S. Calcium detection of human hair and nail by the nanosecond time-gated spectroscopy of laser-ablation plume. In *SPIE 3917, Optical Biopsy II, Proceedings of the BiOS 2000 The International Symposium on Biomedical Optics, San Jose, CA, USA, 22 January 2000*; SPIE Press: Bellingham, DC, USA; pp. 87–92. [CrossRef]
22. Corsi, M.; Cristoforetti, G.; Hidalgo, M.; Legnaioli, S.; Palleschi, V.; Salvetti, A.; Tognoni, E.; Vallebona, C. Application of laser-induced breakdown spectroscopy technique to hair tissue mineral analysis. *Appl. Opt.* **2003**, *42*, 6133–6137. [CrossRef]
23. Emara, E.M.; Imam, H.; Hassan, M.A.; Elnaby, S.H. Biological application of laser induced breakdown spectroscopy technique for determination of trace elements in hair. *Talanta* **2013**, *117*, 176–183. [CrossRef]
24. Zhang, S.; Chu, Y.; Ma, S.; Chen, F.; Zhang, D.; Hu, Z.; Zhang, Z.; Jin, H.; Guo, L. Highly accurate determination of Zn and Cu in human hair by ultrasound-assisted alkali dissolution combined with laser-induced breakdown spectroscopy. *Microchem. J.* **2020**, *157*, 105018. [CrossRef]
25. Araki, S.; Hirai, S.; Mitou, A. Quantification of elements in hair standard data by neutron activation analysis. *Bunseki Kagaku (Anal. Chem.)* **1994**, *43*, 845–850. (In Japanese)
26. NIST: Atomic Spectra Database Lines Form. Available online: https://physics.nist.gov/PhysRefData/ASD/LIBS/libs-form.html (accessed on 15 December 2020).
27. Delaware State University. LIBS Database–Optical Science Center for Applied Research. Available online: https://oscar.desu.edu/libs/ (accessed on 18 January 2021).
28. Farid, N.; Bashir, S.; Mahmood, K. Effect of ambient gas conditions on laser-induced copper plasma and surface morphology. *Phys. Scr.* **2012**, *85*, 015702. [CrossRef]
29. Kyorin Preventive Medicine Institute, Hair Mineral Analysis. Available online: https://kyorin-yobou.net/hair_analysis/ (accessed on 12 December 2020). (In Japanese).
30. Lazic, V.; Fantoni, R.; Colao, F.; Santagata, A.; Morone, A.; Spizzichino, V. Quantitative laser induced breakdown spectroscopy analysis of ancient marbles and corrections for the variability of plasma parameters and of ablation rate. *J. Anal. At. Spectrom.* **2004**, *19*, 429–436. [CrossRef]

Article

Repetitive Detection of Aromatic Hydrocarbon Contaminants with Bioluminescent Bioreporters Attached on Tapered Optical Fiber Elements

Jakub Zajíc [1,2,3,*], Steven Ripp [2], Josef Trögl [4], Gabriela Kuncová [4,5] and Marie Pospíšilová [1]

1 Faculty of Biomedical Engineering, Czech Technical University in Prague, 27201 Kladno, Czech Republic; pospim14@fbmi.cvut.cz
2 Center for Environmental Biotechnology, University of Tennessee, Knoxville, TN 37996, USA; saripp@utk.edu
3 Department of Cardiology, Regional Hospital Liberec, 46063 Liberec, Czech Republic
4 Faculty of Environment, Jan Evangelista Purkyně University in Ústí nad Labem, 40096 Ústí nad Labem, Czech Republic; josef.trogl@ujep.cz (J.T.); kuncova@icpf.cas.cz (G.K.)
5 Institute of Chemical Process Fundamentals of the ASCR, 16502 Prague, Czech Republic
* Correspondence: jakub.zajic@fbmi.cvut.cz

Received: 11 May 2020; Accepted: 4 June 2020; Published: 6 June 2020

Abstract: In this study, we show the repetitive detection of toluene on a tapered optical fiber element (OFE) with an attached layer of *Pseudomonas putida* TVA8 bioluminescent bioreporters. The bioluminescent cell layer was attached on polished quartz modified with (3-aminopropyl)triethoxysilane (APTES). The repeatability of the preparation of the optical probe and its use was demonstrated with five differently shaped OFEs. The intensity of measured bioluminescence was minimally influenced by the OFE shape, possessing transmittances between 1.41% and 5.00%. OFE probes layered with *P. putida* TVA8 were used to monitor liquid toluene over a two-week period. It was demonstrated that OFE probes layered with positively induced *P. putida* TVA8 bioreporters were reliable detectors of toluene. A toluene concentration of 26.5 mg/L was detected after <30 min after immersion of the probe in the toluene solution. Additional experiments also immobilized constitutively bioluminescent cells of *E. coli* 652T7, on OFEs with polyethyleneimine (PEI). These OFEs were repetitively induced with Lauria-Bertani (LB) nutrient medium. Bioluminescence appeared 15 minutes after immersion of the OFE in LB. A change in pH from 7 to 6 resulted in a decrease in bioluminescence that was not restored following additional nutrient inductions at pH 7. The *E. coli* 652T7 OFE probe was therefore sensitive to negative influences but could not be repetitively used.

Keywords: whole-cell biosensor; bioluminescent bioreporter; optical fiber biosensor; toluene; *Pseudomonas putida* TVA8; *Escherichia coli* 652T7

1. Introduction

Primarily controlled chemicals identified in regulations (i.e., United Nations Environment Programme) are selected due to their toxicity in low concentrations, bioaccumulation potential, persistency, carcinogenicity, and repeated assessment in monitoring programs [1,2]. Due to the widespread use of petroleum products and their chemical properties, the most common and stressed water pollutants are benzene, toluene, ethylbenzene and the xylene isomers (BTEX) [3]. It has been shown that these chemicals have adverse health effects, specifically leukemia, cancer, plastic anemia, or bone-marrow disorders in humans, even at low doses [4–6]. Solid-phase microextraction techniques and stir-bar sorptive extraction in combination with chromatographic and mass spectrometric analysis have been widely accepted for the analysis of various water pollutants [7,8].

Bioluminescent bioreporters producing light as a response to the presence of specific pollution, such as organic compounds or metals, have been constructed since 1992 [9]. The development and application of whole cell bioreporters has shown great promise in the laboratory and under controlled conditions [10–12]. In recent years, smartphones [13] and drones [7] have been used as whole cell biosensor devices with bioluminescent bioreporters. Regardless of many assays, case studies, and constructions of unique biosensor devices such as bioluminescent bioreporter integrated circuits (BBICs) [8], there is no company or institution that has yet established the commercial utilization of bioluminescent bioreporters for the evaluation of pollutant bioavailability. Chemical analysis based on gas chromatography and mass spectroscopy (GC/MS) remains the preferred method, although its combination with bioluminescent bioreporters might reduce the cost and increase the speed of identifying polluted sites, interpreting the hazard of pollutants, and predicting suitability for biodegradation as well as monitoring of biodegradation. A main reason for the limited use of bioluminescent bioreporters is legislative regulation against the environmental release of genetically manipulated cells, e.g., [14]. To better establish the potential utility of bioluminescent bioreporters and their complementary optical biosensor devices, more empirical evidence of field applications and studies of reproducibility and stability, as well as facilitation of analytical protocols and biosensor constructions, are needed.

Continuous monitoring with whole cell biosensors requires repeated inoculation [7] or immobilization [15] of cells on the sensing element. In comparison with other immobilization techniques, the immobilization of bioluminescent bioreporters by attachment on a surface of the sensing element has two advantages: the formation of a layer of cells attached to a surface does not involve any dropping or printing machines, and the cells adhere tightly to the sensing element to minimize loss of the detected bioluminescence signal.

Reporter cells immobilized on the tip of an optical fiber can enable continuous measurements in small volume samples and remote localities. The tiny dimensions of an optical fiber (diameter of fiber end <600 μm) make it possible to immobilize only a few cells on the fiber tip. However, such a small number of cells provides a low light intensity, resulting in low biosensor sensitivity. To increase the signal intensity, the light coupling efficiency can be increased by etching the fiber tip and increasing the number of adhered reporter cells by encapsulation into alginate beads [16]. An increase in bioluminescence signal intensity has also been achieved by the immobilization of reporter cells on the wider end (Ø 1 mm–1 cm) of a tapered optical fiber element [17,18].

In our previous research [19], we demonstrated bioluminescent monitoring of toluene over a period of 135 days by adherence of *P. putida* TVA8 to the chemically modified wider end of a tapered optical fiber element (OFE). This study aims to further elucidate the characteristics of such an OFE-type biosensor. Five different OFEs and two bacterial bioluminescent bioreporter strains were examined. Apart from *P. putida* TVA8, whose bioluminescence is positively induced by toluene [20,21], we also immobilized *Escherichia coli* 652T7, whose bioluminescence is constitutive and decreases in the presence of biotoxicants or other factors that affect cell viability [22]. The bioluminescence of cell layers of *P. putida* TVA8, adhered to several types of OFEs, were induced daily with toluene. Similarly, adhered cell layers of *E. coli* 652T7 were exposed to Luria-Bertani (LB) nutritional media for two weeks. Transmission profiles of all OFEs were calculated with a software script [17] and the results were compared to measured light intensities.

2. Materials and Methods

2.1. Materials and Solutions

All chemicals used were purchased from Fisher Scientific. Piranha solution contained concentrated H_2SO_4 and 30% H_2O_2 in a volume ratio of 7:3. LB medium contained tryptone (10 g), yeast extract (5 g) and NaCl (10 g) dissolved in 1 L of distilled water. For solid medium, 17 g of agar was added. Selective LB medium was supplemented with kanamycin (LB_{kan}) at a final concentration of 50 mg/L. Phosphate buffer (PB) contained $Na_2HPO_4 \cdot 12H_20$ (23.637 g) and KH_2PO_4 (8.98 g) in 1 L of distilled water. The trace element solution contained H_3BO_3 (0.062 g) in 1 L of 1M HCl, than $CaCl_2$ (2.94 g), $ZnSO_4 \cdot 7H_2O$ (1.44 g), $CuSO_4 \cdot 5H_2O$ (0.39 g), $Na_2MoO_4 \cdot 2H_2O$ (0.53 g), $MnSO_4 \cdot H_2O$ (3.5 g), and $FeCl_3 \cdot 6H_2O$ (5.4 g/L) was added in 1 L of distilled water. The mineral salt medium (MSM) consisted of $MgSO_4 \cdot 7H_2O$ (0.2 g), NH_4NO_3 (0.2 g), trace element solution (0.1 mL), ferric chloride solution (0.1 mL), PB (100 mL) and distilled water (900 mL). The yeast minimal medium (YMM) consisted of yeast nitrogen base without amino acids (6.7 g/L), synthetic drop-out supplement Y1774 (1.46 g/L), and glucose to the final concentration of 2% (*w*/*v*). For solid medium, 20 g of agar was added.

The toluene induction solution consisted of MSM (19 mL) and toluene-saturated water (1 mL). The final concentration of toluene was 26.5 mg/L, pH 7.2.

The LB induction solution contained 75% of MSM and 25% of LB medium.

The LB, PB and MSM media were autoclaved at 121 °C for 40 min. The trace elements, ferric chloride, and glucose solutions were sterilized by filtration through polytetrafluoroethylene Nalge Nunc Syringe Filters, 0.2 µm from Fischersci.com (Pittsburgh, PA, USA). The YMM was autoclaved at 121 °C for 20 min and was supplemented with sterile glucose to achieve final 2% concentration.

2.2. Microorganisms and Their Cultivation

Bioluminescent bioreporter microorganisms were kindly provided by the University of Tennessee. *Pseudomonas putida* TVA8 is a bioluminescent bioreporter harboring a chromosomal *tod-luxCDABE* fusion [23] and produces bioluminescence in the presence of BTEX (benzene, toluene, ethylbenzene and xylene) and trichlorethylene. The constitutively bioluminescent bacterium *Escherichia coli* 652T7 is a *luxCDABE*-based strain that was used by Du et al. (2015) to monitor the biotoxicity of cellulose nanocrystals [22].

P. putida TVA8 and *E. coli* 652T7 cells were separately cultivated on LB_{kan} agar for 48 h at 28 °C and then reinoculated to LB_{kan} broth. The cultures were placed in a shaking incubator at 100 rpm and 28 °C and grown overnight to an optical density at 600 nm (OD_{600}) of 0.3 ± 0.15 [18,22].

2.3. Tapered Optical Fiber Elements (OFEs)

An OFE manufacturing starts with a preform which is heated in a furnace and drawn into an optical fiber. The narrowing part between the dripped part and the fiber was used as an OFE (Figure S1). The resulting shape is determined by preform size, drawing temperature and weight. The OFEs were kindly donated from the Institute of Photonics and Electronics of the Czech Academy of Sciences. The ends of OFEs were polished. Each OFE was characterized by diameters (D_x) measured in 10 mm distances along its length ($L = Z_{max}$) (Figures 1 and 2). For the purpose of software calculations (transmittance), the OFE shapes were approximated by a bi-exponential equation ("*Exp2*") in Matlab software (Table 1). This software was used by Kalabova et al. (2018) [17] to compare the model simulations to the real bioluminescence measured with OFEs.

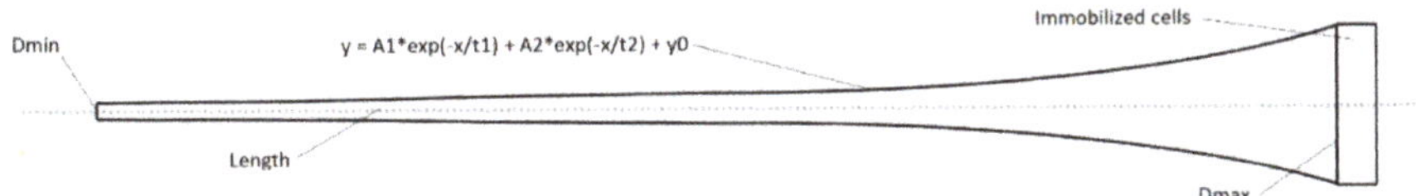

Figure 1. Diagrammatic representation of a tapered optical fiber element (OFE).

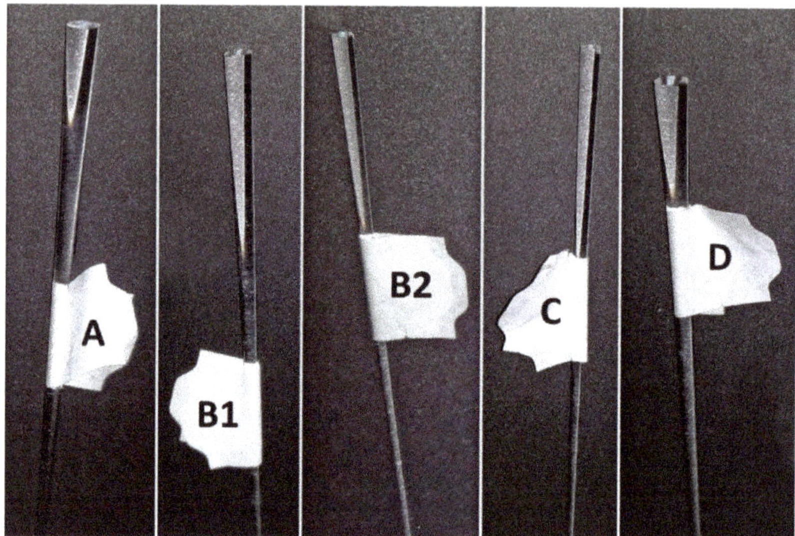

Figure 2. Photographs displaying the wider ends of the OFEs used in this study.

Table 1. Designation of OFEs used in this study. OFEs were approximated using the bi-exponential equation y = y0 + A1*exp(-x/t1) + A2*exp(-x/t2). OFE diameters (*Length, Dmax, Dmin*) and equation parameters (*A1, t1, A2, t2, y0* = 0) are listed.

OFE	Length [mm]	Dmax [mm]	Dmin [mm]	A1	t1	A2	t2
A	328.5	4.97	0.84	3.093	52.826	1.73	429.369
B1	424	4.1	0.73	3.117	69.541	0.984	1501.727
B2	268	4.1	0.89	3.259	73.314	0.832	12963.443
C	532.5	3.04	0.5	1.789	55.036	1.289	554.939
D	208.8	4.85	1.13	1.606	29.886	3.197	186.254

2.3.1. Preparation of OFEs for Chemical Modification

The OFEs were washed in acetone and rinsed with deionized water. The wider end of the OFE was immersed in Piranha solution at 70 °C for 30 min, washed again in deionized water, and dried at 110 °C for 1 h. After drying, the wider ends of the OFEs were modified with (3-aminopropyl)triethoxysilane (APTES) or polyethyleneimine (PEI).

2.3.2. Surface Modification of OFE with APTES

The wider end of each OFE was immersed in a solution of 5% (v/v) APTES in dry toluene at ambient temperature for 24 h. Afterwards, the OFE was rinsed with toluene and acetone and finally dried at 110 °C for 1 h (modified protocol from [24]).

2.3.3. Surface Modification of OFE with PEI

The wider end of the OFE was immersed in a 0.2% (w/v) solution of PEI in deionized water for 30 min and then air-dried [25].

2.4. Adsorption of Bioreporter Cells on OFE Modified Surfaces

The APTES-modified end of the OFE was fixed vertically in an Erlenmeyer flask containing 150 mL of LB_{kan} medium to which was added 1 mL of an overnight culture of either *P. putida* TVA8 or

E. coli 652T7. Cells grew and adsorbed on the wider element end in a shaker at 50 rpm and 28 °C for 4 days (i.e., the minimal time needed for *P. putida* TVA8 cells to be adsorbed [19]).

To increase adsorption of *E. coli* 652T7 to the APTES modified OFE surface, $FeCl_3$ was added to the LB_{kan} growth medium at a final concentration of 150 μM [26].

Due to inadequate growth of *E. coli* 652T7 on the APTES modified OFE surface, the PEI modified surface was attempted as an alternative. A 20 mL aliquot of an overnight culture of *E. coli* 652T7 in LB_{kan} medium was centrifuged at 3000 g for 5 min. The pellet was resuspended in 20 mL MSM and centrifuged again at 3000 g for 5 min. The pellet was then resuspended in 20 mL of 0.2% PEI in MSM and left in a shaker for 30 min at 100 rpm and 28 °C. Finally, the culture was centrifuged at 3000 g for 5 min and the pellet resuspended in 20 mL of MSM [25]. The wider end of the PEI modified OFE was then immersed in the 20 mL suspension of *E. coli* 652T7 and shaken at 50 rpm for 30 min and 28 °C.

2.5. Measurement of Induced Bioluminescence

The thin end of the OFE was attached to a light guiding cable, which was connected to the Oriel 70680 photon multiplier tube. The accelerating voltage of the photon multiplier tube was set to 850 V and the electrical current was manually read from the Oriel 7070 detection system. Experiments were set-up and performed in a light-tight box. The wider end of the OFE with the adsorbed cells was fixed 4 ± 1 mm from the bottom of a 50 mL glass beaker that was then filled with 10 mL of induction solution (toluene for *P. putida* TVA8 or LB/MSM for *E. coli* 652T7), thereby immersing the wider end of the OFE in the induction solution (Figure 3). Reflective aluminum foil was placed underneath the beaker. The current, proportional to the intensity of bioluminescence, was recorded every 30–60 min for 18 h at an ambient temperature of 21 °C. Every 24 h, the wider end of the OFE was gently washed with MSM using a pipette, and then re-immersed into fresh induction solution. Exceptions to these 24 h washings were the 1–3 day pauses for holidays and weekends. On these days, the element remained immersed in the induction solution for up to 72 h. Five different OFEs were tested (Figure 2) with each one being measured over a period of 14–20 days.

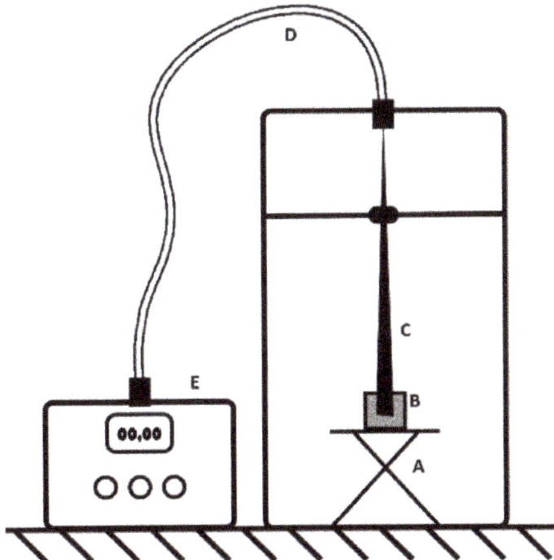

Figure 3. Experimental set-up for monitoring bioluminescence from *P. putida* TVA8 and *E. coli* 652T7 bioreporters adhered to surface modified OFEs. All experiments were performed in a light-tight box that contained an adjustable stand (A), the induction solution (B), the OFE (C), and a light guiding cable (D) that terminated to a photon multiplier measurement device (E).

2.6. Visualization of OFE Cell Adherence Using Scanning Electron Microscopy (SEM)

SEM was used to verify the attachment of *P. putida* TVA8 to its APTES modified OFE. Since the OFEs themselves could not be processed for SEM imaging, quartz cones were used as an alternative (Figure S2). Quartz cones were surface modified with APTES and *P. putida* TVA8 cells were immobilized as explained above. Surface modified quartz cones with adhered cells were placed in a 50 mL beaker containing 30 mL of toluene induction solution. Bioluminescent signaling by the cells was verified by taking light measurements in a Perkin-Elmer IVIS Lumina K imaging system. After two days of immersion, SEM imaging was performed. The quartz cones with immobilized cells were fixed in McDowell–Trump Fixative (Fischer Scientific), gold coated (SPI Module Sputter Coater), and then viewed in a Zeiss Auriga SEM.

2.7. Statistics

Five different OFEs were tested (Figure 2). Five OFEs were modified with APES and one was modified with PEI. Each OFE was used once in a single experiment, which lasted 20 days, where light readings were taken every 30–60 min for 18 h. Replication of the immobilization technique on APTES modified quarz surface, and measurement of bioluminescence was shown. Data from the five OFEs plus one OFE from reference [19] were combined. Bioluminescence maxima, peak integrals, and times of the first bioluminescence maxima within the 20 days were plotted and approximated with polynomial and exponential curves respectively.

3. Results and Discussion

3.1. Imobilization and Induction of Bioluminescence from P. putida TVA8 on APTES Modified OFEs

The APTES modification of OFE led to the successful adherence of *P. putida* TVA8 on its surface. Two days after beginning the immobilization procedure, lumps of cell clusters (100–1000 μm apart) among much smaller scattered clusters or single cells were observed under SEM (Figure 4), which corresponds to standard biofilm establishment characteristics [27]. A fully developed biofilm layer of *P. putida* TVA8 was photographed after 130 days on APTES modified quartz fiber and was presented in a previous paper [19].

The time-records of daily inductions of the five different OFEs are shown in Figure 5.

The intensities of the detected light were low during the first few days after induction and then gradually increased (Figure 6). This might be ascribed to an advanced covering of the base of the OFE with cells. Most time-record curves show two peaks.

The times of the first bioluminescence maxima decreased from more than 10 h to 2–5 h after the fifth day of induction. The same times of the first maxima were observed in a previous study (Figure 7) [19]. The second bioluminescence maxima (appearing after 12–14 h) were probably caused by the bioluminescence of cells growing in the induction solution, which supports the video record (Video S1). This accelerated video shows the course of the induction of *P. putida* TVA8 adhered on the OFE cone. A gradual increase in bioluminescence of cells adhering on the APTES modified surface, the base, and the part of the cone, was followed by a high bioluminescence intensity located only on the base. The bioluminescence of these attached cells fell below the detection limits and at the end (12–15 h) a low light signal emerged in the induction solution.

The measured intensities of bioluminescence significantly differed among the five OFEs. Table S1 compares time-records, bioluminescence maxima, and integrals of bioluminescence. OFEs were repeatedly induced over 20 days and bioluminescence always increased after immersion in the toluene induction solution (A 10x, B1 14x, B2 16x, C 15x, D 14x). To use such an OFE biosensor for the detection of toluene, a signal above twice the background noise of the detector (2×0.26 nA at 850 V) can reliably confirm the presence of toluene in the liquid sample. This level of bioluminescence generation appeared within 0.5 h after immersion in the induction solution. This growth was observed in all inductions (for all OFEs) with exceptions over the first two days. During this initial period, the cell layers were likely not matured and performed slowly, with low bioluminescent responses.

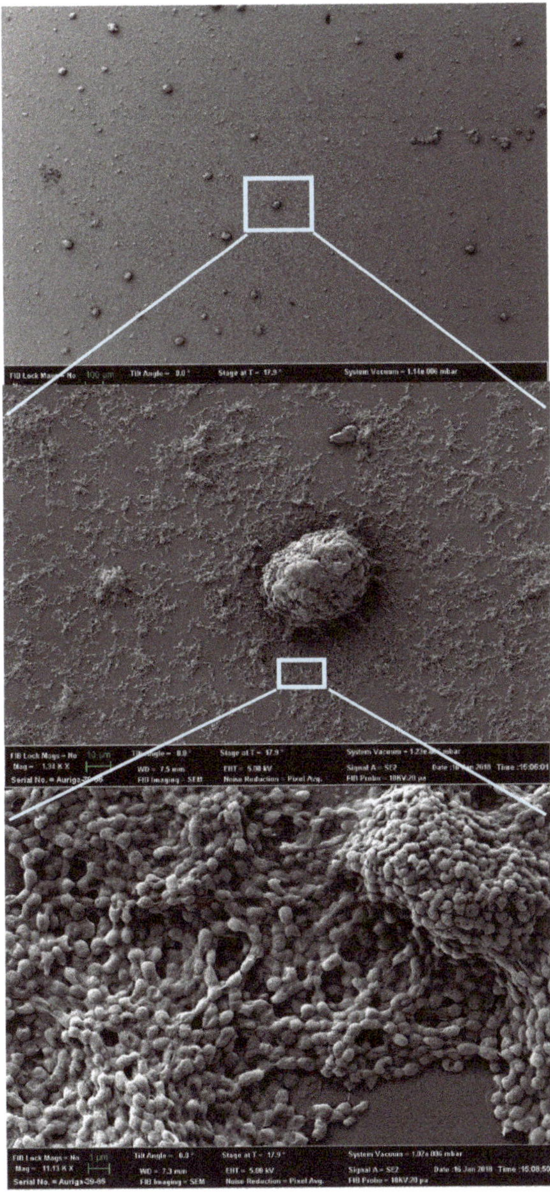

Figure 4. Scanning Electron Microscopy (SEM) image of *P. putida* TVA8 on the (3-aminopropyl) (APTES) modified OFE after the second day of adsorption.

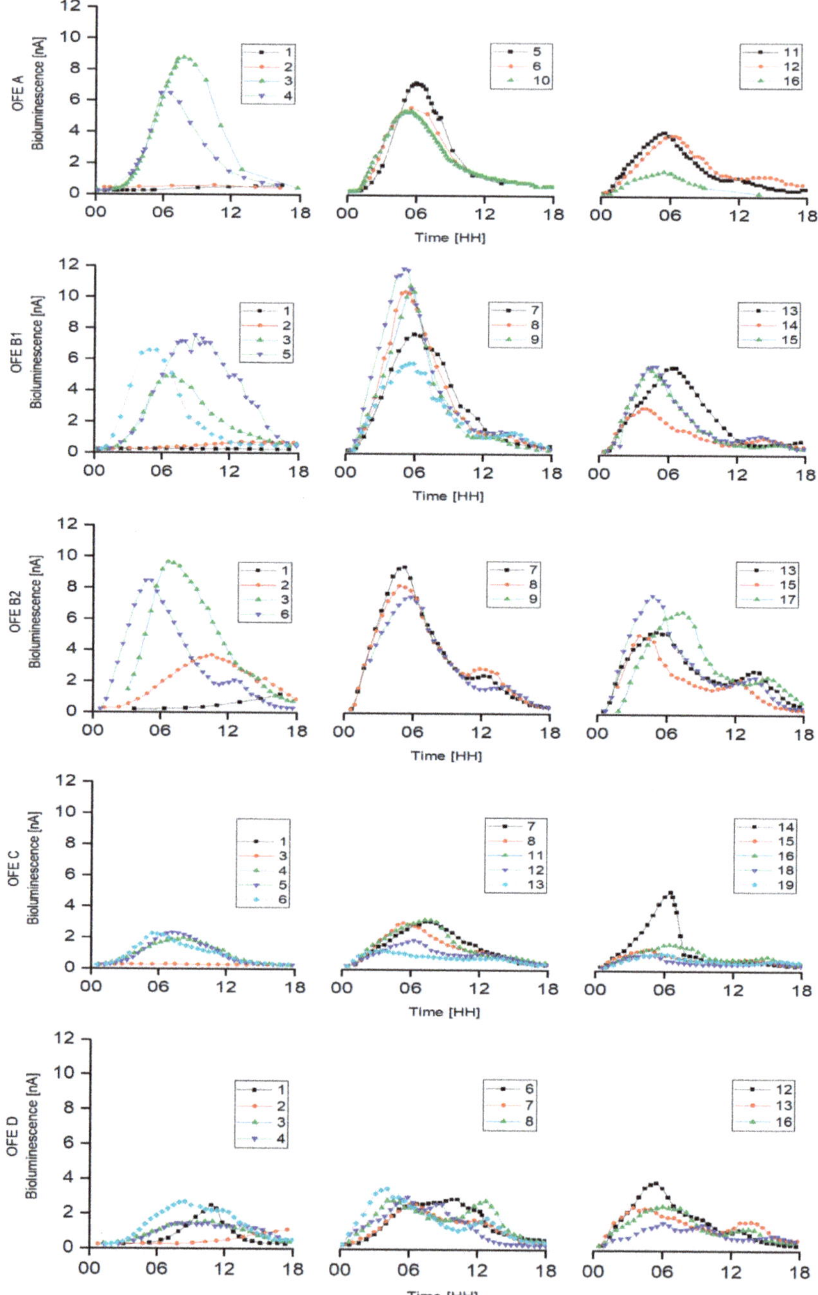

Figure 5. Time-records of daily inductions of bioluminescence of *P. putida* TVA8 immobilized on each of the five available OFEs (A, B1, B2, C, D). Background noise of 26 nA was subtracted from the measured data. *Y*-axis denotes detected bioluminescence intensity in nA. *X*-axis denotes time from the induction of bioluminescence in hours. Each experiment lasted 15–19 days. Chart legends denote the measurement day number.

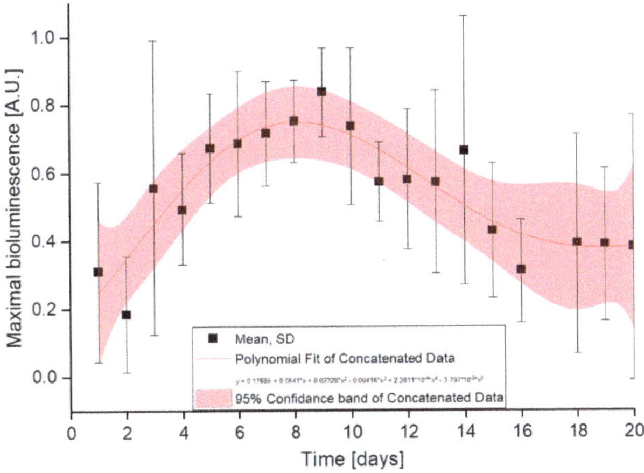

Figure 6. Daily bioluminescence maxima normalized to 1. Aggregated data from the five OFEs plus one OFE from reference [19].

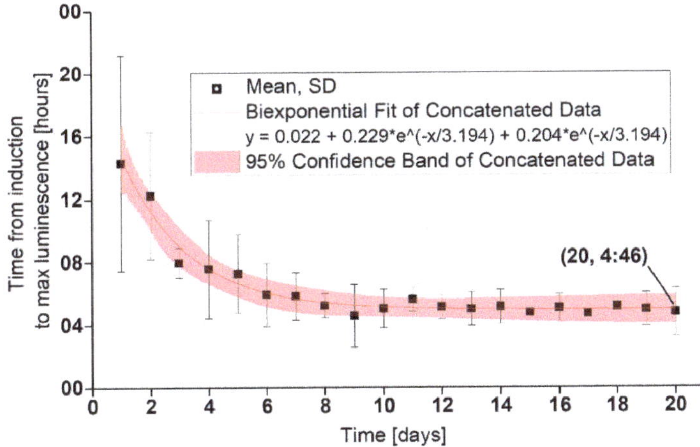

Figure 7. Time of the first bioluminescence maxima. Aggregated data from the five OFEs plus one OFE from reference [19].

These results confirmed previous observations that OFEs adhering with *P. putida* TVA8 can be repeatedly used as a detector for toluene after a few days of the stabilization of immobilized cells [19]. A stabilization period of two to four days was observed, even if the cells were immobilized in silica gel [18,28]. Possible ways to improve the stability of bioluminescence signal responses include engineering a cell strain with two reporter genes (one under control of an analyte of interest and one constitutively present to monitor cell viability). An alternative to this is the use of two independent bioreporters (constitutive and inducible) derived from the same strain. Nevertheless, in the optical fiber arrangement, this resolution requires two fibers, which complicates the sensor construction. The analyte-specific signal must be than corrected according to cell viability [29]; or genetical manipulation of a bacterial strains ability to create and dissolve biofilm structure [30,31].

3.2. Influence of the Shape of OFE

The calculated characteristics of OFEs are presented in Table 2. Transmittances were calculated numerically as a percentage of rays that pass through the OFE from the wider end to a detector, which is placed on the thin end of the OFE. The number of cells on the wider end of each OFE was determined based on the assumption that the cells are spheres, with diameters of 1 μm, organized in one layer [17]. A product of transmittance and number of cells, referred to as OFE efficiency, is the relative amount of light transmitted by the OFE from a monolayer of cells on the wider end to the detector connected to the OFE thin end [32].

Table 2. Calculated transmittance, cell number and efficiency of the five OFEs used in this study.

OFE	Transmittance [%]	Number of Cells × 10^7	OFE Efficiency × 10^5
A	1.41	2.47	3.48
B1	1.62	1.68	2.72
B2	2.21	1.68	3.71
C	1.51	0.92	1.38
D	5.02	2.35	11.75

Using the same mathematical model Kalabova et al. [17], we predicted an outcome of experiments with an OFE and a plastic-clad-silica (PCS) fiber, in this study transmittances calculated among the five different OFEs differed by 3.6% at most (see Table 2). These transmittances are negligible in comparison to the changes in bioluminescence production, likely due to a fluctuation in the number of immobilized cells and their physiological state. Intensities of measured bioluminescence (Figure 5) were independent of OFE transmittances and OFE efficiencies. Intensity of bioluminescence production is susceptible to small variations of temperature, pH, medium composition, bioavailability of inducer to each single cell, and many other factors that cannot be completely controlled and increase the higher variability of the bioluminescence. Calculated transmittances of OFEs significantly increased as their shape moved closer to a frustum cone (Figures S3 and S4). In reality, OFEs always exhibit such a curved shape.

3.3. Immobilization and Induction of Bioluminescence from E. coli 652T7

In the LB_{Kan} growth medium, *E. coli* 652T7 did not adhere on the APTES modified base of the OFEs regardless of the addition of ferric chloride which was added to theoretically enhance the adhesion of microorganisms by lowering the repulsion forces [26]. Cell attachment was observed only at the interface of growth medium and air (Figure 8). At this interface, photon-OFE binding efficiency is <1%, thus bioluminescence of these attached cells did not significantly contribute to detected light.

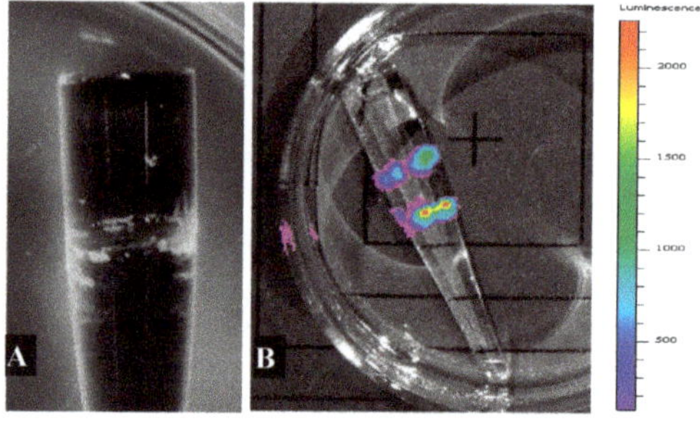

Figure 8. (**A**) *E. coli* 652T7 on the surface of an APTES modified quartz cone after four days in Lauria-Bertani $(LB)_{Kan}$ cultivation medium. (**B**) Bioluminescence of *E. coli* 652T7 induced after immersion in LB medium as measured in an IVIS Lumina K imager (photons/sec/cm^2/steradian)

E. coli 652T7 was immobilized on the base of OFE-D with PEI. The time records of daily inductions with LB medium is presented in Figure 9. Other than the first induction intensities, bioluminescence increased within 15 min after immersing the OFE into the LB solution. The intensities remained stable for 18 h on the first day, 6–9 h on all other days, and then sharply decreased due to a depletion of nutrients. To test the OFE with immobilized *E. coli* 652T7 as a biosensor for biotoxicity, HCl was added to the induction solution on the eighth day. This caused pH lowering to pH = 6 and decreased the bioluminescence, which did not recover after the following two inductions. These results imply that an OFE immobilized with *E. coli* 652T7 is sensitive to influences that affect cell viability but cannot be repetitively used as a biosensor since the cells are dying and not recovering.

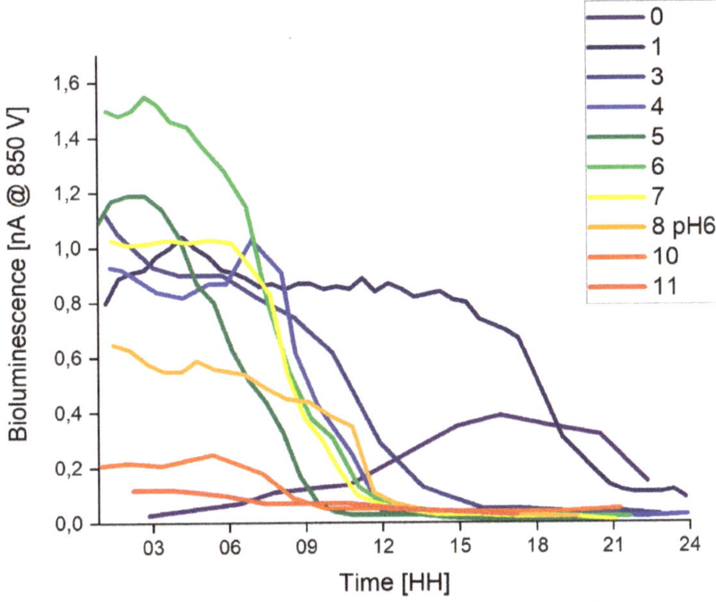

Figure 9. Time records of bioluminescence of *E. coli* 652T7 immobilized on OFE D in polyethyleneimine (PEI). Legend denotes the days after the immobilization.

4. Conclusions

In this study, we immobilized the bioluminescent bioreporter *P. putida* TVA8 on a tapered OFE in order to prepare a biosensor for the detection of liquid toluene. This study broadened our previous research, where we used physico-chemical models, using contact angles and zeta potential, to facilitate the attachment of *P. putida* TVA8 to quartz surfaces after treatment with APTES. The biofilm development of *P. putida* TVA8 with time was quantified and the repeatability of the biofilm preparation and the repeatability of bioluminescence detection was determined. Other than a short maturation period (~5 days), the OFEs exhibited a stable bioluminescent response for at least 20 days.

We additionally immobilized the constitutively bioluminescent toxicity bioreporter *E. coli* 652T7 on a PEI modified OFE and demonstrated its potential use as a biosensor for cytotoxicity. Additionally, the immobilization process that we used, without any bulky matrix requirements, could be applied towards many other microbial bioreporters for the biosensing of a variety of different analytes. However, since the reproducibility of the bioreporter responses remains low, the developed biosensor can be used for online, rapid and multiplexed monitoring of the presence of a pollutant, but not its concentration.

Supplementary Materials: The following are available online at http://www.mdpi.com/1424-8220/20/11/3237/s1. Figure S1: Scheme of OFEs preparation with description; Figure S2: Quartz cones; Video S1: Video record of bioluminescence of Pseudomonas putida TVA8 attached on quartz cone. Video recorded for 17 h and accelerated to 20 s; Table S1: Integrals of intensities and maxima of bioluminescence for five OFEs; Figure S3: Plotted geometrical shape of five theoretical OFEs with identical Dmax, Dmin, (radius r) and length L, but different shape (bent), OFE models were used to calculate their transmittances; Figure S4: Calculated transmittances of the five theoretical OFEs based on their geometrical shapes.

Author Contributions: Conceptualization, J.Z., M.P. and S.R.; methodology, J.Z., S.R.; validation, J.Z., G.K. and S.R.; formal analysis, J.Z., J.T., M.P., G.K.; investigation, J.Z.; resources, J.Z., S.R.; data curation, J.Z. and G.K..; writing—original draft preparation, J.Z.; writing—review and editing, G.K., J.T. and S.R.; visualization, J.Z..; supervision, S.R. and M.P.; funding acquisition, J.Z. and J.T. All authors have read and agreed to the published version of the manuscript.

Funding: This research was funded by the internal Student Grant Competition of the Czech Technical University in Prague, CZ, grant numbers SGS17/110/OHK4/1T/17 and SGS18/097/OHK4/1T/17 and by Ministry of Education, Youth and Sports of the Czech Republic via project Smart City–Smart Region–Smart Community, grant number CZ.02.1.01/0.0/0.0/17_048/0007435.

Acknowledgments: Workplace, material support, and microorganisms were kindly provided by the University of Tennessee, Knoxville, TN, USA.

Conflicts of Interest: The authors declare no conflict of interest.

References

1. Jones, O.; Preston, M.R.; Fawell, J.; Mayes, W.; Cartmell, E.; Pollard, S.; Harrison, R.M.; Mackenzie, A.R.; Williams, M.; Maynard, R.; et al. *Pollution: Causes, Effects and Control*; Royal Society of Chemistry: London, UK, 2015.
2. Hill, M.K. *Understanding Environmental Pollution*; Cambridge University Press: Cambridge, UK, 2010.
3. Mitra, S.; Roy, P. BTEX: A Serious Ground-water Contaminant. *Res. J. Environ. Sci.* **2011**, *5*, 394–398. [CrossRef]
4. Aksoy, M. Benzene and Leukemia. *Environ. Health Perspect.* **1991**, *91*, 165. [CrossRef] [PubMed]
5. Lynge, E.; Andersen, A.; Nilsson, R.; Barlow, L.; Pukkala, E.; Nordlinder, R.; Boffetta, P.; Grandjean, P.; Heikkiia, P.; Horte, L.-G.; et al. Risk of cancer and exposure to gasoline vapors. *Am. J. Epidemiol.* **1997**, *145*, 449–458. [CrossRef] [PubMed]
6. Lim, S.K.; Shin, H.S.; Yoon, K.S.; Kwack, S.J.; Um, Y.M.; Hyeon, J.H.; Kwak, H.M.; Kim, J.Y.; Kim, T.H.; Kim, Y.J.; et al. Risk assessment of volatile organic compounds benzene, toluene, ethylbenzene, and xylene (BTEX) in consumer products. *J. Toxicol. Environ. Heal. Part A* **2014**, *77*, 1502–1521. [CrossRef]
7. Lu, Y.; Macias, D.; Dean, Z.S.; Kreger, N.R.; Wong, P.K. A UAV-Mounted Whole Cell Biosensor System for Environmental Monitoring Applications. In Proceedings of the IEEE Transactions on Nanobioscience, Quebec, QC, Canada, 12 November 2015; pp. 811–817.
8. Nivens, D.E.; McKnight, T.E.; Moser, S.A.; Osbourn, S.J.; Simpson, M.L.; Sayler, G.S. Bioluminescent bioreporter integrated circuits: Potentially small, rugged and inexpensive whole-cell biosensors for remote environmental monitoring. *J. Appl. Microbiol.* **2004**, *96*, 33–46. [CrossRef]
9. Heitzer, A.; Webb, O.F.; Thonnard, J.E.; Sayler, G.S. Specific and quantitative assessment of naphthalene and salicylate bioavailability by using a bioluminescent catabolic reporter bacterium. *Appl. Environ. Microbiol.* **1992**, *58*, 1839–1846. [CrossRef]
10. Diplock, E.E.; Alhadrami, H.A.; Paton, G.I. Commercial Application of Bioluminescence Full Cell Bioreporters for Environmental Diagnostics. In *Handbook of Hydrocarbon and Lipid Microbiology*; Springer: Berlin, Heidelberg, 2010; pp. 4445–4458.
11. Close, D.M.; Ripp, S.; Sayler, G.S. Reporter Proteins in Whole-Cell Optical Bioreporter Detection Systems, Biosensor Integrations, and Biosensing Applications. *Sensors* **2009**, *9*, 9147–9174. [CrossRef]
12. Gutiérrez, J.C.; Amaro, F.; Martín-González, A. Heavy metal whole-cell biosensors using eukaryotic microorganisms: An updated critical review. *Front. Microbiol.* **2015**, *6*. [CrossRef]
13. Kim, H.; Jung, Y.; Doh, I.J.; Lozano-Mahecha, R.A.; Applegate, B.; Bae, E. Smartphone-based low light detection for bioluminescence application. *Sci. Rep.* **2017**, *7*, 40203. [CrossRef]

14. Regional Environmental Center and Umweltbundesamt GmbH, Chemicals and GMO Legislation. *Handbook on the Implementation of EC Environmental Legislation*; European Union: Szentendre, Hungary, 2014; Section 8; pp. 93–126.
15. Lobsiger, N.; Stark, W.J. Strategies of immobilizing cells in whole-cell microbial biosensor devices targeted for analytical field applications. *Anal. Sci.* **2019**, *35*, 839–847. [CrossRef]
16. Polyak, B.; Bassis, E.; Novodvorets, A.; Belkin, S.; Marks, R.S. Bioluminescent whole cell optical fiber sensor to genotoxicants: System optimization. *Sens. Actuators B. Chem.* **2001**, *74*, 18–26. [CrossRef]
17. Kalabova, H.; Pospisilova, M.; Jirina, M.; Kuncova, G. Mathematical Model for Laboratory System of Bioluminescent Whole-Cell Biosensor with Optical Element. *J. Biosens. Bioelectron.* **2018**, *9*, 1–5.
18. Kuncova, G.; Ishizaki, T.; Solovyev, A.; Trogl, J.; Ripp, S. The Repetitive Detection of Toluene with Bioluminescence Bioreporter Pseudomonas putida TVA8 Encapsulated in Silica Hydrogel on an Optical Fiber. *Materials* **2016**, *9*, 467. [CrossRef] [PubMed]
19. Zajic, J.; Bittner, M.; Branyik, T.; Solovyev, A.; Sabata, S.; Kuncova, G.; Pospisilova, M. Repetitive inductions of bioluminescence of Pseudomonas putida TVA8 immobilised by adsorption on optical fiber. *Chem. Pap.* **2016**, *70*, 877–887. [CrossRef]
20. Kuncova, G.; Pazlarova, J.; Hlavata, A.; Ripp, S.; Sayler, G.S. Bioluminescent bioreporter Pseudomonas putida TVA8 as a detector of water pollution. Operational conditions and selectivity of free cells sensor. *Ecol. Indic.* **2011**, *11*, 882–887. [CrossRef]
21. Applegate, B.M.; Kehrmeyer, S.R.; Sayler, G.S. A chromosomally based tod-luxCDABE whole-cell reporter for benzene, toluene, ethybenzene, and xylene (BTEX) sensing. *Appl. Environ. Microbiol.* **1998**, *64*, 2730–2735. [CrossRef]
22. Du, L.Y.; Arnholt, K.; Ripp, S.; Sayler, G.; Wang, S.Q.; Liang, C.H.; Wang, J.K.; Zhuang, J. Biological toxicity of cellulose nanocrystals (CNCs) against the luxCDABE-based bioluminescent bioreporter Escherichia coli 652T7. *Ecotoxicology* **2015**, *24*, 2049–2053. [CrossRef]
23. Sanseverino, J.; Applegate, B.M.; King, J.M.H.; Sayler, G.S. Plasmid-mediated mineralization of naphthalene, phenanthrene, and anthracene. *Appl. Environ. Microbiol.* **1993**, *59*, 1931–1937. [CrossRef]
24. Wang, J.; Du, S.; Onodera, T.; Yatabe, R.; Tanaka, M.; Okochi, M.; Toko, K. An SPR sensor chip based on peptide-modified single-walled carbon nanotubes with enhanced sensitivity and selectivity in the detection of 2,4,6-trinitrotoluene explosives. *Sensors* **2018**, *18*, 4461. [CrossRef]
25. D'Souza, S.F.; Melo, J.S.; Deshpande, A.; Nadkarni, G.B. Immobilization of yeast cells by adhesion to glass surface using polyethylenimine. *Biotechnol. Lett.* **1986**, *8*, 643–648. [CrossRef]
26. Fletcher, M. Attachment of Pseudomonas fluorescens to glass and influence of electrolytes on bacterium-substratum separation distance. *J. Bacteriol.* **1988**, *170*, 2027–2030. [CrossRef] [PubMed]
27. Kokare, C.R.; Chakraborty, S.; Khopade, A.N.; Mahadik, K.R. Biofilm: Importance and applications. *Indian. J. Biotechnol.* **2009**, *8*, 159–168.
28. Kuncova, G.; Pospisilova, M.; Solovyev, A. Optical Fiber Whole Cell Bioluminescent Sensor. In Proceedings of the XX International Conference on Bioencapsulation, Orillia, ON, Canada, 21–24 September 2012; B. Abstr. 2012, proceedings reference P_01. pp. 96–97.
29. Roda, A.; Roda, B.; Cevenini, L.; Michelini, E.; Mezzanotte, L.; Reschiglian, P.; Hakkila, K.; Virta, M. Analytical strategies for improving the robustness and reproducibility of bioluminescent microbial bioreporters. *Anal. Bioanal. Chem.* **2011**, *401*, 201–211. [CrossRef]
30. Angelaalincy, M.J.; Navanietha Krishnaraj, R.; Shakambari, G.; Ashokkumar, B.; Kathiresan, S.; Varalakshmi, P. Biofilm Engineering Approaches for Improving the Performance of Microbial Fuel Cells and Bioelectrochemical Systems. *Front. Energy Res.* **2018**, *6*. [CrossRef]
31. Wei, Q.; Ma, L.Z. Biofilm matrix and its regulation in Pseudomonas aeruginosa. *Int. J. Mol. Sci.* **2013**, *14*, 20983–21005. [CrossRef]
32. Pospisilova, M.; Kuncova, G.; Trogl, J. Fiber-Optic Chemical Sensors and Fiber-Optic Bio-Sensors. *Sensors* **2015**, *15*, 25208–25259. [CrossRef]

 sensors

MDPI

Article

Cultivation of *Saccharomyces cerevisiae* with Feedback Regulation of Glucose Concentration Controlled by Optical Fiber Glucose Sensor

Lucie Koštejnová [1,*], Jakub Ondráček [1], Petra Majerová [1], Martin Koštejn [1], Gabriela Kuncová [1,2] and Josef Trögl [2]

1 Institute of Chemical Process Fundamentals of the CAS, v. v. i., Rozvojová 135/1, 16502 Prague, Czech Republic; ondracek@icpf.cas.cz (J.O.); majerova@icpf.cas.cz (P.M.); kostejn@icpf.cas.cz (M.K.); kuncova@icpf.cas.cz (G.K.)

2 Faculty of Environment, Jan Evangelista Purkyně University in Ústí nad Labem, Pasteurova 3632/15, 40096 Ústí nad Labem, Czech Republic; Josef.Trogl@ujep.cz

* Correspondence: kostejnova@icpf.cas.cz; Tel.: +420-220-390-303

Abstract: Glucose belongs among the most important substances in both physiology and industry. Current food and biotechnology praxis emphasizes its on-line continuous monitoring and regulation. These provoke increasing demand for systems, which enable fast detection and regulation of deviations from desired glucose concentration. We demonstrated control of glucose concentration by feedback regulation equipped with in situ optical fiber glucose sensor. The sensitive layer of the sensor comprises oxygen-dependent ruthenium complex and preimmobilized glucose oxidase both entrapped in organic–inorganic polymer ORMOCER®. The sensor was placed in the laboratory bioreactor (volume 5 L) to demonstrate both regulations: the control of low levels of glucose concentrations (0.4 and 0.1 mM) and maintenance of the glucose concentration (between 2 and 3.5 mM) during stationary phase of cultivation of *Saccharomyces cerevisiae*. Response times did not exceed 6 min (average 4 min) with average deviation of 4%. Due to these regulation characteristics together with durable and long-lasting (≥2 month) sensitive layer, this feedback regulation system might find applications in various biotechnological processes such as production of low glucose content beverages.

Keywords: yeast cultivation; feedback regulation; glucose detection; optical biosensor

Citation: Koštejnová, L.; Ondráček, J.; Majerová, P.; Koštejn, M.; Kuncová, G.; Trögl, J. Cultivation of *Saccharomyces cerevisiae* with Feedback Regulation of Glucose Concentration Controlled by Optical Fiber Glucose Sensor. *Sensors* **2021**, *21*, 565. https://doi.org/10.3390/s21020565

Received: 4 December 2020
Accepted: 11 January 2021
Published: 14 January 2021

1. Introduction

The number of people with diabetes has risen from 108 million in 1980 to 422 million in 2014, and between 2000 and 2016 there was a 5% increase in premature mortality from diabetes the World Health Organization (WHO) predicts that the diabetes will be the seventh leading cause of death in 2030 [1]. The accurate evaluation of the glucose content in foods is extremely important for the maintenance of its physiological level in blood of diabetic individuals. Information about glucose content of foods and beverages is essential for both producers and consumers. Glucose monitoring is crucial in tracing the fermentation processes in the wine, brewing, and dairy industries.

The first qualitative test of glucose was published in 1848 [2]. Since that time many methods of glucose quantification have been described [3]. Web of Science links to 25,000 references for key words glucose detection. Plenty of physical detection principles have been used which are often non-specific to glucose [4]. For example, microwave resonator-based sensors might be advantageous for glucose detection in blood [5,6] or in some industrial applications as their linear range of measured glucose concentrations is from zero to more than ten weight percent [7]. In comparison with these microwave-based sensors, optical biosensors with glucose oxidase exhibit a high specificity to glucose.

The first glucose biosensor was realized in 1962 using glucose oxidase and a Clark electrode [8]. Since that time, hundreds of glucose biosensors have been described with electrochemical or optical transducers [9]. Prevailing research effort was focused on biosensors for monitoring concentration of glucose in blood and other physiological fluids [10,11]. Nevertheless, glucose biosensors for beverages and food industry have also been presented. Ayenimo et al. described a polypyrrole-based bilayer amperometric glucose biosensor integrated with a permselective layer, which was successfully employed for glucose determination in various fruit juices [12]. For glucose and galactose detection in fruit juices and skim milk, the graphite working electrode, on which glucose oxidase and β-galactosidase were coimmobilized by means of covalent bonding, was developed by Portaccio et al. [13]. Amperometric biosensor based on modified screen-printed carbon electrodes for online glucose monitoring during cultivation of *Saccharomyces cerevisiae* in microbioreactor was published by Panjan et al. [14]. Otten at al. described a fluorescence resonance energy transfer (FRET)-based glucose biosensor, which can be applied in microbioreactor-based cultivations. The soluble sensor was successfully applied online to monitor the glucose concentration in an *Escherichia coli* culture [15].

Optical fiber sensors have showed a number of advantages over electrochemical sensors due to their independence of electromagnetic involvement, security, sensitivity, ruggedness, and fine dimensions of probes. In the last twenty years, the progress in the development of optical fiber sensors of glucose was reviewed by Wolfbeis [16–20] and Wang [21–23]. Among various principles of optical fiber sensing of glucose, enzymatic sensors with glucose oxidase and optical oxygen transducers have been the most broadly studied and used. The combination of high selectivity of glucose oxidase and ruggedness of optical oxygen transducer allow them to be applied in industrial processes.

Except for fast and precise detection, food and biotechnology processes require low deviations from desired glucose concentration ($c_{GL}{}^{DES}$) and their quick compensation. An advantageous solution to these demands is an incorporation of biosensoric detection components into feedback loops, which keep actual glucose concentration (c_{GL}) on desired value ($c_{GL} = c_{GL}{}^{DES}$), or in allowed limits throughout production process. A continuous measurement of glucose in real-time without connection to glucose dispenser was described by Maldonado et al. [24] and Blankenstein et al. [25].

The system of control of glucose concentrations of a perfusion medium in a rotating wall perfused vessel bioreactor culturing BHK-21 cells was presented by Xu et al. in 2004. The custom-made glucose sensor was based on a hydrogen peroxide electrode. The system first controlled the glucose concentration in perfusing medium between 4.2 and 5.6 mM for 36 days and then at different glucose levels for 19 days. A stock solution with a high glucose concentration (266 mM) was used as the glucose injection solution. The standard error of prediction for glucose measurement by the sensor, compared to measurement by the Beckman glucose analyzer, was 0.4 mM for 55 days [26]. Commercially available systems of glucose control in bioreactors (CITSens Bio, SEG-Flow) use the TRACE filtration probe for harvesting cell-free filtrate from bioreactors and fermenters under sterile conditions. Company Stratophase Ltd. (Hampshire, United Kingdom) developed a sensor that regulated glucose concentration with an in situ optical glucose sensor. The sensor with Bragg grating measures glucose concentrations as changes of refractive index. Such measurements are nonspecific and, therefore, they suffer from great errors due to changes of concentrations of the other medium components.

Enzymatic glucose sensor with oxygen transducer, its preparation and analytical features, were described in previous papers [27–29]. The sensor withstands sterilization by UV and ethanol as well as mechanical stresses caused by mixing of fermentation broth. The sensor is based on the measurement of oxygen consumption due to oxidation of glucose catalyzed by an enzyme, glucose oxidase. Ruthenium complex serves as an optical transducer. Its fluorescence is quenched proportionally with oxygen concentration. Both sensitive parts—the enzyme (preimmobilized on Sepabeads®) and the complex—are coated in Ormocer® (organic–inorganic polymer). The polymer has a siloxane network and is UV curable. Covalent attachment to a carrier (Sepabeads®) protects the enzyme against harsh conditions after mixing with Ormocer® and during the UV curing of the sensitive layer on the acrylate lens.

Here, we present on-line feedback regulation of glucose concentration controlled by this glucose sensor placed in the bioreactor vessel. The regulation was demonstrated in both modes: in dilution and fed batch cultivation. Dilution mode was proposed for production of beverages with limited content of glucose and for diabetics, where glucose content should be close to zero. In fed-batch cultivation, microorganisms consume glucose, but to keep cells alive and producing, the concentration of glucose must be maintained at a level which allows cells to survive but limits their proliferation. In the fed batch process, glucose regulation is demonstrated with the most industrially used microorganism, *Saccharomyces cerevisiae.*

2. Materials and Methods

2.1. Chemicals

D-glucose p.a., sucrose p.a, glutaraldehyde 25%, $K_2HPO_4 \cdots 3\ H_2O$ p.a., and KH_2PO_4 p.a. were purchased from Penta s.r.o. (Czechia). Yeast extract, glucose oxidase type X-S from *Aspergilus niger* with specific activity 228.4 kU/g ($GOX^{X\text{-}S}$), and glucose oxidase type II-S from *Aspergilus niger* with specific activity 37.7 kU/g ($GOX^{II\text{-}S}$) were from Sigma Aldrich® (St. Louis, MO, USA). Bacteriological pepton was purchased from Oxoid-Thermo Fisher Scientific Inc. (Waltham, MA, USA). NaOH p.a. and H_3PO_4 85% were purchased from Lach-Ner s.r.o. (Prague, Czechia). Sepabeads® EC-HA 403 (SEPA) were delivered by Resindion S.r.l (Binasco, Italy). Tris (4,7-difenyl-1,10-fenantrolin) ruthenium(II) dichloride (RuC) was purchased from ABCR GmbH (Karlsruhe, Germany). Ormocer® KSK 1248 (ORM) was obtained from Fraunhofer Institute for Silicate Research ISC (Wurzburg, Germany). Photoinitiator Irgacure® 500 was from BASF, Germany. Polymethylmethacrylate biconvex lens with diameter 7 mm was from Institute of Plasma Physics of the CAS, v.v.i. (Prague, Czechia).

2.2. Media and Gasses

Potassium phosphate buffer (50 mM, *pH* 7) was prepared by dilution $K_2HPO_4 \cdots 3\ H_2O$ (21.1 mL) and KH_2PO_4, (28.9 mL) in distilled water (1 L). Final *pH* 7 was adjusted by addition of NaOH (0.2 M) or H_3PO_4 (0.2 M).

YPG cultivation medium contained yeast extract (1 g), peptone (2 g) and glucose (2 g) in distilled water (100 mL).

Concentrated glucose solution (1 mM) was prepared by dilution glucose (90 g) in distilled water (500 mL).

Oxygen, 2.5 UN1072 and nitrogen 4.0, UN1066 were products by Linde Gas a.s. (Prague, Czechia).

2.3. Microorganisms

Saccharomyces cerevisiae was obtained from Collection of microorganisms of the Institute of Biochemistry and Microbiology UCT Prague. Overnight culture (50 mL) was added into the bioreactor. Optical density (*OD*) of the overnight culture (3× diluted) was 0.5.

OD was determined in 1 cm cuvette at 600 nm by UV–VIS spectrophotometer HP8452A (Hewlett-Packard, Palo Alto, CA, USA).

2.4. Preparation of Optical Sensitive Layers

Sensitive layers were prepared by procedure described in details by Kostejnova et al. [29]. Briefly, 200 mg of SEPA was activated by a stirring with glutaraldehyde (4 mL) for four hours. After centrifugation and washing, the enzyme solution was added to activated SEPA and the mixture was stirred for 18 h with the same velocity as was used during activation. The mixture was centrifuged, the supernatant was removed, and Sepabeads® with immobilized glucose oxidase (SEPA-GOX) were washed twice with buffer. Ormocer® was mixed with Ru complex, Irgacure 500 and sucrose to form ORM-RC. Components of sensitive layers were mixed on glass slide in the ratio 2:1, ORM-RC: SEPA-GOX. The mixtures were deposited on plastic lenses and cured by UV light for ten minutes. After UV polymerization the lenses were immersed in phosphate buffer (50 mM, *pH* 5.9) overnight to wash out sucrose. The thicknesses were measured with the microscope Tescan (Czech Republic) in the center and on the periphery of the layers (see Figure 1). Parameters of preparation and analytical characteristics (sensitivity (*SN*), linear dynamic range (*LDR*), and response time (*RT*) of sensitive layers used in the tests are presented in Table 1 and on Figure 1.

Table 1. Parameters and analytical characteristics of sensitive layers.

Test	Velocity of Stirring in Activation of Sepabeads® (rpm)	Enzyme (mg_{enzyme}/g_{SEPA})		Thickness * (nm)	*SN* (μs L $mmol^{-1}$)	*LDR* (mM)	*RT* (min)
		$GOX^{X\text{-}S}$	$GOX^{II\text{-}S}$				
Dilution	20	125	—	280	0.452	0–1.5	3
Cultivation	20	—	12.5	225	0.091	0–7	5
Response reproducibility **	800	125	—	300	0.306	0–1.6	9

* The average of thicknesses measured with the microscope Tescan. ** Average values of *SN*, *LDR*, and *RT* during testing of reproducibility of biosensor response.

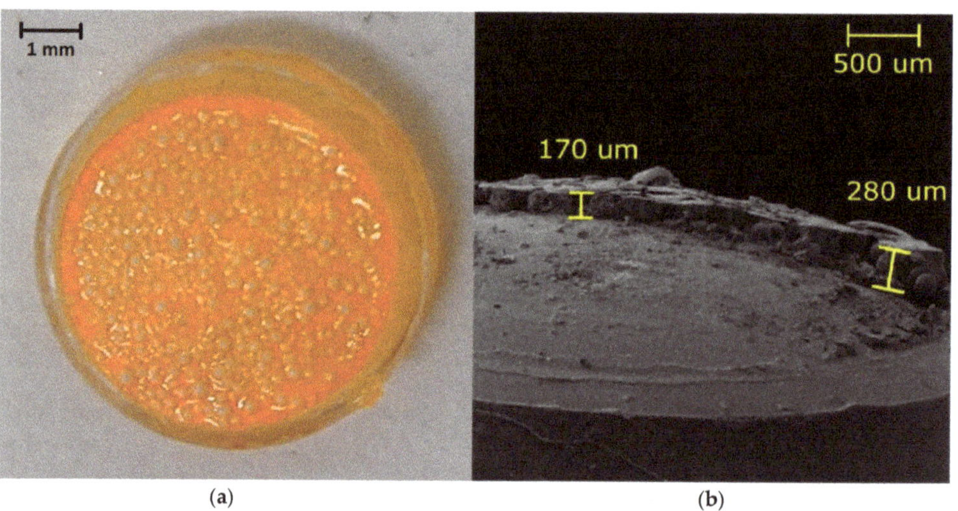

(a) (b)

Figure 1. Photos of optical sensitive layer. (**a**) The lens with sensitive layer (used in the cultivation). (**b**) Scanning electron microscopy photo of cross-section of this sensitive layer.

2.5. Feedback Regulation System

A schema with photos of the feedback regulation system is on Figure 2.

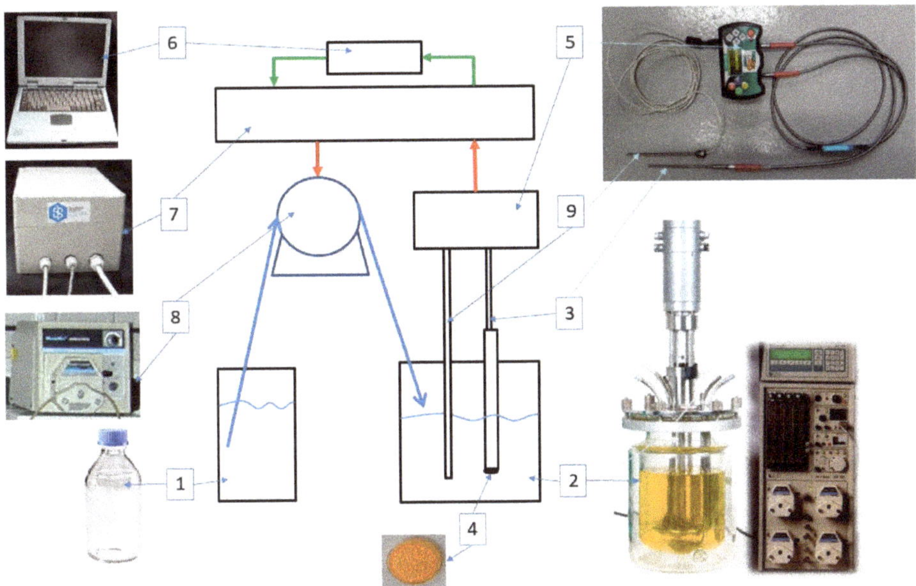

Figure 2. Feedback regulation system. 1. container with buffer or concentrated glucose solution, 2. bioreactor Applicon (5 L), 3. optical fibres, 4. lens with sensitive layer, 5. light source and the detector (λ_{EX} = 470 nm, λ_{EM} = 580 nm), 6. PC LabView, 7. control unit, 8. peristaltic pump, 9. temperature probe of NeoFox.

The bioreactor (Figure 2, position 2) was produced by Applicon V.B. (Schiedam, The Netherland), it is autoclavable, comprised of a tempered glass vessel with volume of 5 L and inner diameter 166 mm equipped with Bio controller ADI 1010 and Bio console ADI 1025 for control of *pH* (pH electrode), the concentration of dissolved oxygen (*dO2*) is measured with Clark electrode, has temperature control (resistance thermometer), and monitors the velocity of mixing.

Optical probe for measurement of glucose concentration had identical shape as *pH* and *dO2* probes of the bioreactor (Figure 2, position 3–4). The probe consisted from stainless steel tube with glucose sensitive layer, and a bundle of optical fibres connected to the light source and the detector (Figure 2, position 5). NeoFox Sport made by Ocean Optic (Largo, FL, USA) was used as light source and the detector. Increase of glucose concentrations in the reactor was detected with sensitive layer on acrylate lens (Figure 2. position 4) as the increase of fluorescence lifetime of RuC due to consumption of oxygen in oxidation of glucose catalyzed by glucose oxidase.

Analog output signal from NeoFox was read by serial port of the control unit (Figure 2. position 8). After evaluation with control software (Figure 2, position 7) the output signal was controlling rotation speed of peristaltic pump Masterflex 7518-00 (Cole-Parmer Instrument Company, Vernon Hills, IL, USA) (Figure 2, position 9). Measurement and control software was developed in Labview software environment. The Labview control software allows for following settings: (1) desired glucose concentration ${c_{GL}}^{DES}$, which was set as life time set point; τ^{DES} (2) a mode of concentration control—diluting or feeding; (3) time resolution of glucose concentration measurement—averaging time, t_{checII}; and (4) speed of glucose dosing with peristaltic pump as percentage of pump power. Actual measured glucose concentration and indication of pump action (run/stop) were displayed on the monitor. The software recorded measured glucose concentrations c_{GL} and course of administration of buffer or concentrated solution of glucose. The system responded on deviation from ${c_{GL}}^{DES}$ after checking time (t_{chec}). During t_{chec} the system only measures without responding. In dilution mode, the pump administrated buffer in case that c_{GL} was the same or higher than maximum allowed glucose concentration ($c_{GL}{}^{MAX}$). In cultivation mode, the concentrated glucose solution was added into the bioreactor in case that c_{GL} was the same or lower than minimum allowed glucose concentration ($c_{GL}{}^{MIN}$). To reach desired concentration, $c_{GL} = c_{GL}{}^{DES}$, the volume of a dose (buffer/conc. glucose solution) was calculated by the software with respect of volume of liquid in the bioreactor. The new t_{chec} started after the pump finished dosing.

2.6. Reproducibility of Biosensor Response in Repetitive Measurements during 2 Months

The probe of the biosensor with sensitive layer (enzyme concentration 125 mg $GOX^{X\text{-}S}$/g SEPA, thickness of the layer 300 nm) was immersed in non-sterile buffer (50 mM, *pH* 7), which was bubbled by sterile air with volume flow 16 mL/min, mixed 400 rpm, and tempered 25 °C. During two months, on working days, *SN*, *LDR*, and *RT*. were determined once a day. Measurements and calculations of analytical characteristics are described in details in previous paper [29].

2.7. Sterilization

The bioreactor filled with medium/buffer with inserted *pH*, *dO2*, *T* probe, together with storage bottles of base, acid, concentrated glucose, dilution buffer, and all connection pipes was sterilized in autoclave at 120 °C for 30 min. Before inserting into bioreactor, glucose probe, and acrylate lens with sensitive layer were sterilized by immersing in ethanol (70%) for 5 min and irradiation with UV for 10 min.

2.8. Off-Line Measurement of Glucose Concentration

Off-line glucose concentration was measured with Glucose oxidase Activity Assay kit from Sigma Aldrich s.r.o. (Prague, Czechia).

2.9. Control of Glucose Concentration with Feedback Regulation System

2.9.1. Feedback Regulation of Glucose Concentration to Lower Level (Dilution Mode).

The bioreactor filled by 2 L buffer (50 mM, *pH* 7) was bubbled by sterile air with volume flow 16 mL/min, mixed 400 rpm, and tempered 25 °C. The value of *dO2* in the bioreactor was 21%. The regulation was tested at two maximum concentrations, $c_{GL}{}^{MAX}$ = 0.5 mM and 0.125 mM for corresponding $c_{GL}{}^{DES}$ = 0.4 mM and $c_{GL}{}^{DES}$ = 0.1 mM, respectively.

NeoFox was switch on and the system was left to stabilize fluorescence lifetime (τ^0) at zero glucose concentration ($c_{GL}{}^{0}$) for 15 min. To calibrate Neofox, glucose concentration was increased by addition of concentrated glucose solution to $c_{GL}{}^{DES}$ = 0.4 mM or 0.1 mM (Figures 3 and 4. green frames), which lead to increase of fluorescence lifetime (τ^{DES}). On user interface monitor were set τ^0, τ^{DES} and corresponding $c_{GL}{}^{0}$, $c_{GL}{}^{DES}$, times for averaging t_{chec} = 10 min and $c_{GL}{}^{MAX}$ = 0.5 mM resp. 0.125 mM (Figures 3 and 4, red frames) and corresponding τ^{MAX} calculated from the calibration. After three t_{chec} (Figures 3 and 4, position 1) concentration of glucose was increased from $c_{GL}{}^{DES}$ = 0.4 mM resp. 0.1 mM to $c_{GL}{}^{MAX}$ = 0.5 mM resp. 0.125 mM by hand pipetting of solution of concentrated glucose (0.2 mL, resp. 0.05 mL) into the bioreactor (Figures 3 and 4, position 2). After t_{chec}, $c_{GL}{}^{MAX}$ was detected and the pump of feedback loop dosed calculated buffer volume into the bioreactor to reach $c_{GL} = c_{GL}{}^{DES}$ (Figures 3 and 4, position 3). In reality, glucose concentration after regulation ($c_{GL}{}^{REG}$) differs from $c_{GL}{}^{DES}$. The cycles of addition of glucose solution and regulation were repeated three times during both tests. The test for $c_{GL}{}^{DES}$ = 0.4 mM was reproduced three times (Figure 3.I–III).

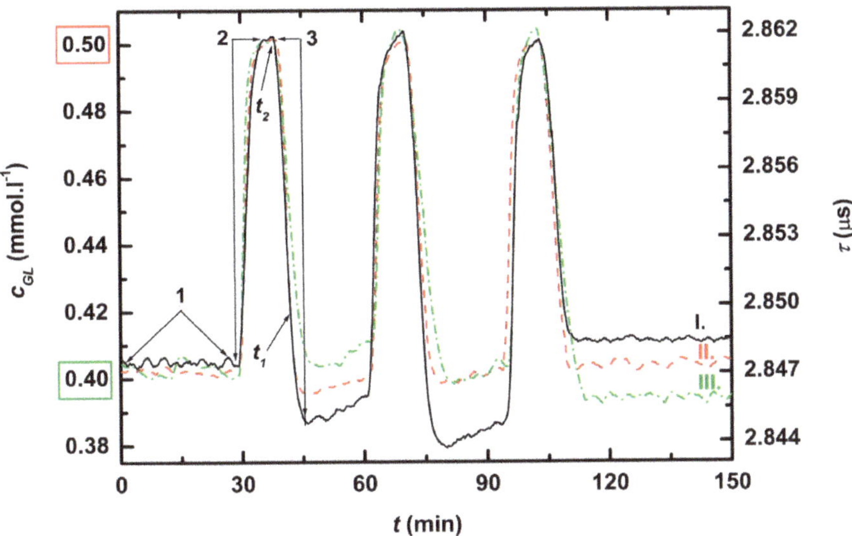

Figure 3. Time record of the feedback regulation of lower glucose concentration (dilution with buffer) for three independent experiments at the same conditions (I.-III.). After the initial three t_{chec} (position 1), the concentration of glucose was increased from $c_{GL}{}^{DES}$ = 0.4 mM to $c_{GL}{}^{MAX}$ = 0.5 mM by hand pipetting of concentrated glucose solution into the bioreactor (position 2). After $c_{GL}{}^{MAX}$ detection, the pump of feedback loop dosed buffer into the reactor (position 3). t_1 is time when actual measured glucose concentration exceeded glucose concentration after regulation for 10% and t_2 is time when buffer was added.

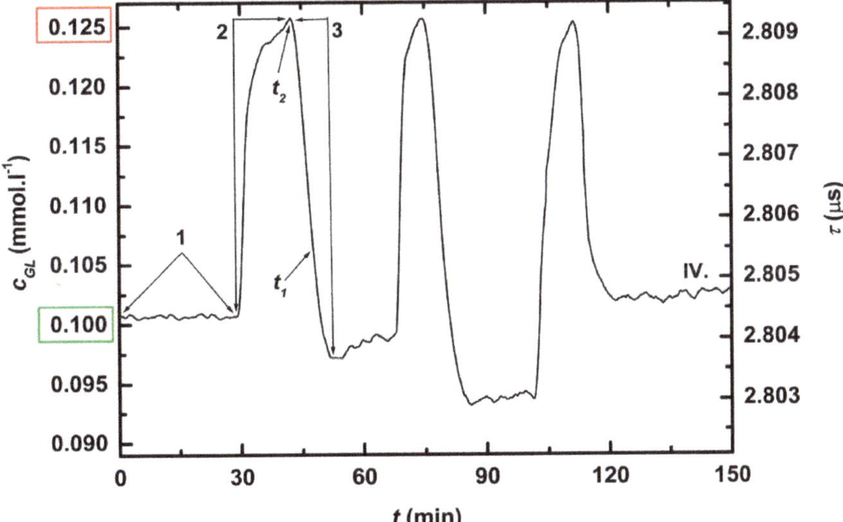

Figure 4. Time record of the feedback regulation of lower glucose concentration (dilution with buffer, test IV). After the initial three t_{chec} (position 1), the concentration of glucose increased from $c_{GL}{}^{DES}$ = 0.100 mM to $c_{GL}{}^{MAX}$ = 0.125 mM by hand pipetting of concentrated glucose solution into the bioreactor (position 2). After $c_{GL}{}^{MAX}$ detection, the pump of feedback loop dosed buffer into the reactor (position 3). t_1 is the time when actual measured glucose concentration exceeded glucose concentration after regulation for 10% and t_2 is time when buffer was added.

Response time (RT_{90}) was calculated for each buffer dose according to equation

$$RT_{90} = t_1 - t_2, \tag{1}$$

where t_1 is time when actual measured glucose concentration exceeded glucose concentration after regulation for 10% $c_{GL} = c_{GLREG} + 0.1 \times c_{GLREG}$, and t_2 is time when buffer was added.

Deviation (s) of glucose concentration after regulation from $c_{GL}{}^{DES}$ was calculated for each buffer dose

$$s = ((c_{GL}{}^{DES} - c_{GL}{}^{REG})/c_{GL}{}^{DES}) \times 100. \tag{2}$$

2.9.2. Feedback Regulation of Glucose Concentration of Fed Batch Cultivation of *Saccharomyces cerevisiae* in Stationary Phase (Cultivation Mode)

The bioreactor was filled with 2 L incomplete YPG medium. Throughout the experiment, the bioreactor was tempered to 30 °C and bubbled with sterile air, oxygen, or nitrogen to keep constant $dO2$ = 21%. Concentration of glucose in incomplete YPG medium was measured off-line. Double point glucose calibration was done in the first hour of the experiment. The first point was the concentration of glucose in incomplete YPG medium $c_{GL}{}^{MIN}$ (2 mM, Figure 4, red frame) and the second point was $c_{GL}{}^{DES}$ = 3.5 mM (Figure 4, green frame) acquired by hand pipetting of concentrated glucose (3 mL). Neofox corresponding fluorescence lifetimes, τ^{MIN} and τ^{DES}, were set on user software monitor together with t_{chec}. Double the response time (RT_{90}), determined in calibration, was opted for checking time, t_{chec} = 10 min.

After calibration, glucose (40 g) was added to complete YPG medium, thus c_{GL} = 111 mM, which was out of the range (0–7 mM) of the biosensor (Figure 5, position 1). The bioreactor was inoculated with night culture of *Saccharomyces cerevisiae* (50 mL, OD for 3x diluted culture was 0.5) and feedback regulation was switched on (Figure 5, position 2). The growing cells consumed glucose. After 11.5 h of fermentation, c_{GL} dropped below 7 mM (Figure 5, position 3). Within 6 t_{chec}, the measured glucose concentration c_{GL} decreased from 6.8 to 1.9 mM and $c_{GL} < c_{GL}{}^{MIN}$ (Figure 6. position 1). At the end of 6th t_{chec} the pump started to dose concentrated glucose solution into the reactor so that $c_{GL} = c_{GL}{}^{DES}$, *resp.* $c_{GL}{}^{REG}$, after the seventh t_{chec} (Figure 6, position 2). Culture of *Saccharomyces cerevisiae* consumed added glucose during the 8th t_{chec} and $c_{GL}{}^{DES}$, and $c_{GL}{}^{REG}$ dropped to (or under) $c_{GL}{}^{MIN}$, which activated dosing pump (Figure 6, position 3). The cycle kept adding glucose to reach $c_{GL}{}^{DES}$ followed by consumption with yeast culture to $c_{GL}{}^{MIN}$ (cycle ↓↑) was repeated seven times. The experiment was twice reproduced.

Response times were calculated for each glucose dose according to

$$RT_{90}{}^{*} = t_1 - t_2, \tag{3}$$

where t_1 is time when measured glucose concentration reach 90% of concentration after regulation: $c_{GL} = 0.9.\ c_{GL}{}^{REG}$ and t_2 is time when concentrated glucose solution was added.

Deviation (s^*) from $c_{GL}{}^{DES}$ were calculated for each glucose dose according to

$$s^* = (c_{GL}{}^{DES} - c_{GL}{}^{REG})/c_{GL}{}^{DES} \times 100. \tag{4}$$

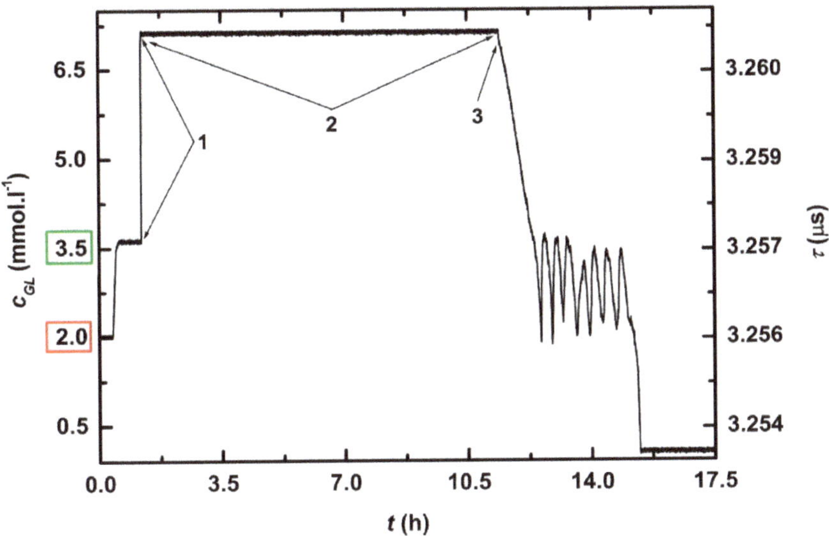

Figure 5. Time record of glucose concentration during fed-batch cultivation of *Saccharomyces cerevisiae*. After two-point calibration ($c_{GL}{}^{DES}$ = 3.5 mM, $c_{GL}{}^{MIN}$ = 2 mM), glucose was added to complete YPG medium (c_{GL} = 111 mM), which was out of the range (0-7 mM) of the biosensor (position 1). The bioreactor was inoculated with an overnight culture of Saccharomyces cerevisiae and the feedback regulation was switched on (position 2). After 11.5 h of fermentation, c_{GL} dropped below 7 mM (position 3).

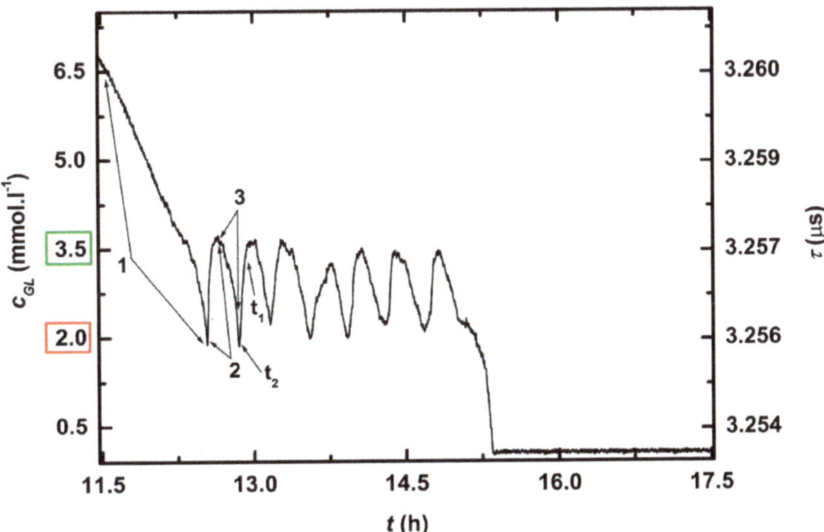

Figure 6. Detailed time record of feedback regulation of glucose concentration during stationary phase of cultivation of *Saccharomyces cerevisiae*.

3. Results and Discussion

3.1. Feedback Regulation of Glucose Concentration to Lower Level (Dilution Mode)

This regulation is demonstration of application of feedback system in production of beverages for diabetes, where the demand is to keep glucose concentration at a level close to zero. The sensitive layer, used in dilution mode, was chosen to meet the need of the lowest detection limit. Based on our previous study [29], such demand best fit sensitive layer comprising high content of enzyme, which is immobilized on undivided SEPA.

Monitoring and control of low glucose concentration levels are in Figures 3 and 4, and characteristics of regulation are in Table 2. In all experiments, response times were shorter than 5 min ($RT_{90} \leq 5$ min). Average deviation was 1.8% for glucose concentration hold at 0.4 mM. For lower glucose concentration, the relative precision of measurement decreased. Therefore, in case the desired glucose concentration was equal to limit of detection (*LOD*) of used biosensor ($c_{GL}{}^{DES}$ = *LOD* = 0.1 mM, the test IV.), the average deviation increased to 3.7%.

Table 2. Response times of the biosensor and deviations from desired glucose concentration in case of step increase of glucose concentration.

Experiment	$c_{GL}{}^{DES}$ (mM)	$c_{GL}{}^{REG}$ (mM)	RT_{90} (min)	s (%)
I.	0.4	0.388	3	3
		0.380	4	5
		0.411	4	3
II.	0.4	0.404	2	1
		0.400	4	0
		0.393	4	2
III.	0.4	0.396	3	1
		0.398	5	1
		0.400	5	0
IV.	0.1	0.097	5	3
		0.093	5	7
		0.101	2	1

3.2. Feedback Regulation of Concentration of Glucose of Fed Batch Cultivation of Saccharomyces Cerevisiae in Stationary Phase (Cultivation Mode)

In cultivation mode, the sensitive layer should possess fast response time and wide concentration range to measure glucose in sufficiently broad concentration range during stationary phase of cultivation. In our previous paper [29], it was shown that *LDR* of the layers increased with decreasing enzyme concentration. It was also shown that crushing of spherical SEPA with immobilized glucose oxidase resulted in higher *LDR*. Unfortunately, increasing *LDR* simultaneously increased *RT*, which is an undesirable effect for the feedback regulation system. Therefore, we must compromise between opposing demands on analytical features of sensitive layer for cultivation mode. We used the sensitive layer with $RT \leq 6$ min and *LDR* = 0–7 mM.

A time record of complete cultivation of *Saccharomyces cerevisiae* is presented on Figure 5 and the detail of stationary phase, while glucose concentration was controlled with the feedback regulation system, is on Figure 6. Table 3 shows that in all seven cycles ↓↑, response times were below 6 min ($RT_{90}{}^{*} \leq 6$ min) and deviation from regulation did not exceed 9%. The average $RT_{90}{}^{*}$ was 4 min and the average deviation 3.9%.

In situ monitoring and control glucose concentration during cultivation were described by Tric et al. [30]. They used also enzymatic sensor with optical glucose transducer; however, glucose oxidase was fixed on optically isolated oxygen sensor with glutaraldehyde and covered by perflorated hydrophilic membrane. In comparison with this report, where response times were 6 min for increasing and 10 min for decreasing of glucose concentrations, we reached response times shorter than 6 min in all tests. The shorter response times might be related to faster diffusion of oxygen and glucose in sensitive layer comprising both enzyme and fluorescent complex in one mixture. These results implicate that regulation response times less than few minutes are hard to reach with enzymatic glucose sensor with optical oxygen transducer. Response times become shorter as the activity of enzyme increases and sensitive layer is thinner [26]. Nevertheless, these parameters are limited by technical feasibility of a preparing such layer. Selectivity, robustness, and long-term reliability are favored features of enzymatic glucose sensors with oxygen transducers for control of glucose concentrations in biotechnological processes but, if one minute or less response times are necessary, another type of glucose sensor should be used.

Activity of microorganisms resulted in c_{GL} decreased from 6.8 to 1.9 mM ($c_{GL} < c_{GLMIN}$, position 1). After this point, the pump started to dose concentrated glucose solution into the reactor so that $c_{GL} = c_{GL}{}^{DES}$ (position 2). Culture of *Saccharomyces cerevisiae* continued to consume glucose and c_{GL} dropped to c_{GLMIN}, which activated dosing pump again (position 3). $t_1{}^*$ is time when measured glucose concentration reached 90% of concentration after regulation and $t_2{}^*$ is time when concentrated glucose solution was added.

Table 3. Response times and deviation from desired glucose concentration during feedback control of glucose concentration in stationary phase of cultivation of *Saccharomyces cerevisae*.

Number of of cycle ↓↑	$c_{GL}{}^{DES}$ (mM)	$c_{GL}{}^{REG}$ (mM)	RT_{90} (min)	s^* (%)
1	3.5	3.7	6	6
2	3.5	3.6	3	3
3	3.5	3.6	3	3
4	3.5	3.2	6	9
5	3.5	3.5	4	0
6	3.5	3.4	2	3
7	3.5	3.4	4	3

Deviation (s^*) defined in Equation (4).

3.3. Reproducibility of the Biosensor Response during 2 Month.

During two months (42 measurements) the average *SN* was 0.306 µs L mmol^{-1} with relative deviation 10% (Figure 7) and an average maximum of linear dynamic range (LDR^{MAX}) 1.6 mM with relative deviation 12% (Figure 8). At the first measurement, *RT* was 9 min. In the second measurement, *RT* increased to 14.7 min and this response time was preserved in following 40 measurements. An average *RT* (without the first day) was 15.1 min with relative determinative deviation 8% (Figure 9) and it remained constant throughout the repetitions ($p > 0.9971$). After the first experiment, an increase of *RT* is probably a result of an adsorption of microorganisms from non-sterile buffer, which cause diffusion slowdown of both substrates glucose and oxygen, in the sensitive layer.

3.4. Wider Applicability of the Biosensor

The presented biosensor was developed with immobilized glucose oxidase aiming for the on-line monitoring of glucose concentration. Together with oxygen and pH, glucose concentration is one of the most often measured parameters in biotechnology. Nevertheless, the presented concept is general and replacing of glucose oxidase by other oxidases can result in various analogical biosensors, such as for biological amines [26] or cholesterol oxidase [27] for use on continuous systems. Of interest in near future might be sensors of various environmental pollutants. Biodegradation pathways of many organic pollutants often start with oxygenases enzymes [28,29] of different specificity, and these could serve as a biosensing elements for regulation of continuous water treatment processes.

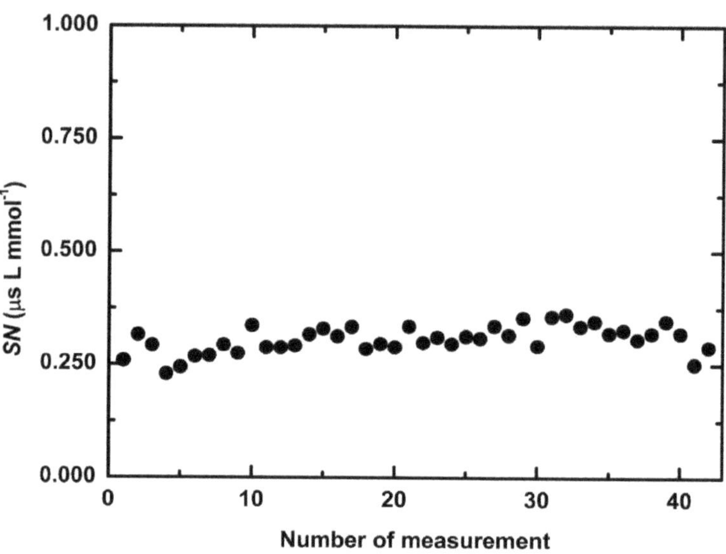

Figure 7. Sensitivity of the optical sensitive layer in 42 repeated measurements during two months.

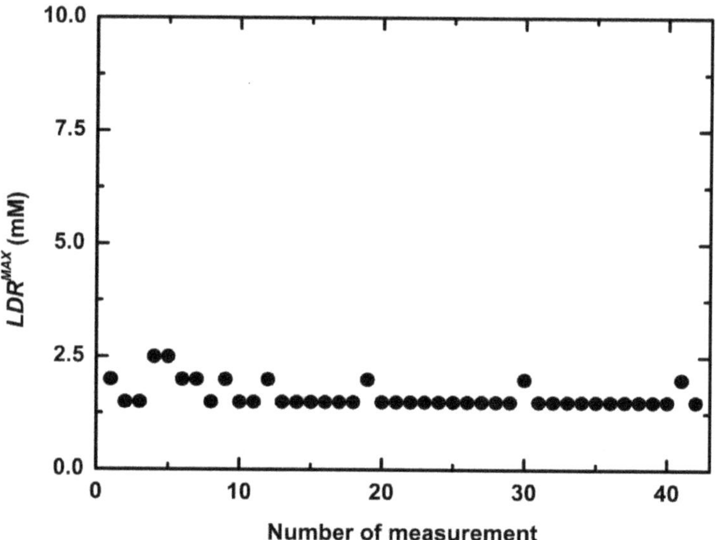

Figure 8. Maxima of linear dynamic range of the optical sensitive layer in 42 repeated measurements during two months.

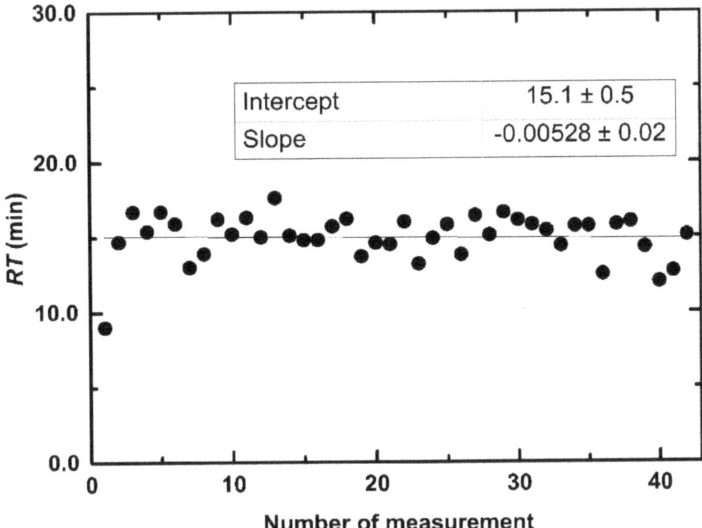

Figure 9. Response time of the optical sensitive layer in 42 repeated measurements during two months.

4. Conclusions

In this work, we presented the feedback system for regulation of glucose concentration based on the enzymatic sensor with optical oxygen transducer. The system was demonstrated for the case of maintaining low glucose concentration. An undesirable increase of glucose concentration was compensated below 0.125 mM by dilution in less than 5 min. In stationary phase of fed batch cultivation when glucose was continuously consumed by growing microorganisms, the feedback system adjusted glucose concentration to 3.5 mM in less than 6 min after detection of the concentration drop to 2 mM. The two-month stability and reproducibility of biosensor response was demonstrated by daily measurements, in which relative determinative deviations of analytical characteristics (sensitivities, linear dynamic ranges, and response times) were less than 12%. In comparison to known and commercially available glucose concentration regulations, the presented feedback system has advantage in use of in situ sensor, robust construction, and long-term stability.

Author Contributions: Conceptualization, L.K. and J.O.; methodology, J.O.; software, J.O.; validation, L.K. and P.M.; formal analysis, L.K. and M.K.; investigation, P.M. and J.O.; resources, G.K.; data curation, L.K.; writing—original draft preparation, L.K.; writing—review and editing, L.K., G.K. and J.T.; visualization, M.K.; supervision, G.K.; project administration, L.K., J.T., and G.K.; funding acquisition, J.T. and G.K. All authors have read and agreed to the published version of the manuscript. Please turn to the CRediT taxonomy for the term explanation. Authorship must be limited to those who have contributed substantially to the work reported.

Funding: The research was co-funded by the Ministry of Youth and Sports via the project Smart City-Smart Region-Smart Community (CZ.02.1.01/0.0/0.0/17_048/0007435).

Data Availability Statement: The data presented in this study are available on request from the corresponding author. The data are not publicly available due to Pat. No. CZ30355 / PUV 2016-33183.

Conflicts of Interest: The authors declare no conflict of interest.

Abbreviations

cycle ↓↑	Cycle adding glucose to reach ${c_{GL}}^{DES}$ followed by consumption with yeast culture to ${c_{GL}}^{MIN}$
τ^{MAX}	Fluorescence lifetime corresponded with maximum allowed glucose concentration
τ^{MIN}	Fluorescence lifetime corresponded with minimum allowed glucose concentration
τ^{DES}	Fluorescence lifetime corresponded with desired glucose concentration
τ^{0}	Fluorescence lifetime corresponded with zero glucose concentration
c_{GL}	Actual glucose concentration
${c_{GL}}^{MAX}$	Maximum allowed glucose concentration
${c_{GL}}^{MIN}$	Minimum allowed glucose concentration
${c_{GL}}^{DES}$	Desired glucose concentration
${c_{GL}}^{REG}$	Glucose concentration after the regulation
${c_{GL}}^{0}$	Zero glucose concentration
dO_2	Concentration of dissolved oxygen
$GOX^{X\text{-}S}$	Glucose oxidase type X-S from *Aspergillus niger* with specific activity 228.4 kU g^{-1}
$GOX^{II\text{-}S}$	Glucose oxidase type II-S from *Aspergillus niger* with specific activity 37.7 kU g^{-1}
LOD	Limit of detection
LDR	Linear dynamic range
LDR^{MAX}	Maximum of linear dynamic range
OD	Optical density
ORM-RC	Mixture prepared by mixing Ormocer® with Ru complex and Irgacure 500
RuC Tris	(4,7-difenyl-1,10-fenantrolin) ruthenium(II) dichloride
RT	Res^{po}nse time of the biosensor during testing of reproducibility of biosensor response
RT_{90}	Response time of biosensor during diluting of solution
${RT_{90}}^{*}$	Response time of biosensor during cultivation
s	Deviation from desired glucose concentration after regulation during diluting of solution
s^{*}	Deviation from desired glucose concentration after regulation during cultivation
SEPA	Sepabeads® EC-HA 403
SEPA-GOX	Sepabeads® with immobilized glucose oxidase
SN	Sensitivity
t_{CHEC}	Checking time

References

1. World Health Organization Global Report on Diabetes. Available online: https://apps.who.int/iris/bitstream/handle/10665/204871/9789241565257_eng.pdf (accessed on 6 January 2021).
2. Clarke, S.F.; Foster, J.R. A History of Blood Glucose Meters and Their Role in Self-Monitoring of Diabetes Mellitus. *Br. J. Biomed. Sci.* **2012**, *69*, 83–93. [CrossRef] [PubMed]
3. Galant, A.L.; Kaufman, R.C.; Wilson, J.D. Glucose: Detection and Analysis. *Food Chem.* **2015**, *188*, 149–160. [CrossRef] [PubMed]
4. Mehrotra, P.; Chatterjee, B.; Sen, S. EM-Wave Biosensors: A Review of RF, Microwave, Mm-Wave and Optical Sensing. *Sensors* **2019**, *19*, 1013. [CrossRef] [PubMed]
5. Yilmaz, T.; Foster, R.; Hao, Y. Radio-Frequency and Microwave Techniques for Non-Invasive Measurement of Blood Glucose Levels. *Diagnostics* **2019**, *9*, 6. [CrossRef] [PubMed]
6. Omer, A.E.; Shaker, G.; Safavi-Naeini, S.; Kokabi, H.; Alquié, G.; Deshours, F.; Shubair, R.M. Low-Cost Portable Microwave Sensor for Non-Invasive Monitoring of Blood Glucose Level: Novel Design Utilizing a Four-Cell CSRR Hexagonal Configuration. *Sci. Rep.* **2020**, *10*, 15200. [CrossRef] [PubMed]
7. Juan, C.G.; Bronchalo, E.; Potelon, B.; Quendo, C.; Avila-Navarro, E.; Sabater-Navarro, J.M. Concentration Measurement of Microliter-Volume Water–Glucose Solutions Using Q Factor of Microwave Sensors. *IEEE Trans. Instrum. Meas.* **2019**, *68*, 2621–2634. [CrossRef]
8. Clark, L.C.; Lyons, C. Electrode Systems for Continuous Monitoring in Cardiovascular Surgery. *Ann. N. Y. Acad. Sci.* **1962**, *102*, 29–45. [CrossRef]
9. Sabu, C.; Henna, T.K.; Raphey, V.R.; Nivitha, K.P.; Pramod, K. Advanced Biosensors for Glucose and Insulin. *Biosens. Bioelectron.* **2019**, *141*, 111201. [CrossRef]
10. Gutiérrez-Capitán, M.; Baldi, A.; Fernández-Sánchez, C. Electrochemical Paper-Based Biosensor Devices for Rapid Detection of Biomarkers. *Sensors* **2020**, *20*, 967. [CrossRef]
11. Teymourian, H.; Barfidokht, A.; Wang, J. Electrochemical Glucose Sensors in Diabetes Management: An Updated Review (2010–2020). *Chem. Soc. Rev.* **2020**, *49*, 7671–7709. [CrossRef]
12. Ayenimo, J.G.; Adeloju, S.B. Amperometric Detection of Glucose in Fruit Juices with Polypyrrole-Based Biosensor with an Integrated Permselective Layer for Exclusion of Interferences. *Food Chem.* **2017**, *229*, 127–135. [CrossRef] [PubMed]

13. Portaccio, M.; Lepore, M. Determination of Different Saccharides Concentration by Means of a Multienzymes Amperometric Biosensor. *J. Sens.* **2017**, *2017*, 1–8. [CrossRef]
14. Panjan, P.; Virtanen, V.; Sesay, A.M. Towards Microbioprocess Control: An Inexpensive 3D Printed Microbioreactor with Integrated Online Real-Time Glucose Monitoring. *Analyst* **2018**, *143*, 3926–3933. [CrossRef] [PubMed]
15. Otten, J. A FRET-Based Biosensor for the Quantification of Glucose in Culture Supernatants of ML Scale Microbial Cultivations. *Microb. Cell Factories* **2019**, *18*, 1–10. [CrossRef]
16. Wolfbeis, O.S. Fiber-Optic Chemical Sensors and Biosensors. *Anal. Chem.* **2000**, *72*, 81–90. [CrossRef]
17. Wolfbeis, O.S. Fiber-Optic Chemical Sensors and Biosensors. *Anal. Chem.* **2002**, *74*, 2663–2678. [CrossRef]
18. Wolfbeis, O.S. Fiber-Optic Chemical Sensors and Biosensors. *Anal. Chem.* **2004**, *76*, 3269–3284. [CrossRef]
19. Wolfbeis, O.S. Fiber-Optic Chemical Sensors and Biosensors. *Anal. Chem.* **2006**, *78*, 3859–3874. [CrossRef]
20. Wolfbeis, O.S. Fiber-Optic Chemical Sensors and Biosensors. *Anal. Chem.* **2008**, *80*, 4269–4283. [CrossRef]
21. Wang, X.-D.; Wolfbeis, O.S. Fiber-Optic Chemical Sensors and Biosensors (2008−2012). *Anal. Chem.* **2013**, *85*, 487–508. [CrossRef]
22. Wang, X.; Wolfbeis, O.S. Fiber-Optic Chemical Sensors and Biosensors (2013–2015). *Anal. Chem.* **2016**, *88*, 203–227. [CrossRef] [PubMed]
23. Wang, X.; Wolfbeis, O.S. Fiber-Optic Chemical Sensors and Biosensors (2015–2019). *Anal. Chem.* **2020**, *92*, 397–430. [CrossRef] [PubMed]
24. Lladó Maldonado, S.; Panjan, P.; Sun, S.; Rasch, D.; Sesay, A.M.; Mayr, T.; Krull, R. A Fully Online Sensor-Equipped, Disposable Multiphase Microbioreactor as a Screening Platform for Biotechnological Applications. *Biotechnol. Bioeng.* **2019**, *116*, 65–75. [CrossRef]
25. Blankenstein, G.; Spohn, U.; Preuschoff, F.; Thommes, J.; MR, K. Multichannel Flow-Injection-Analysis Biosensor System for on-Line Monitoring of Glucose, Lactate, Glutamine, Glutamate and Ammonia in Animal Cell Culture. *Biotechnol. Appl. Biochem.* **1994**, *20*, 291–307. [CrossRef] [PubMed]
26. Xu, Y.; Sun, J.; Mathew, G.; Jeevarajan, A.S.; Anderson, M.M. Continuous Glucose Monitoring and Control in a Rotating Wall Perfused Bioreactor. *Biotechnol. Bioeng.* **2004**, *87*, 473–477. [CrossRef] [PubMed]
27. Rose, K.; Dzyadevych, S.; Fernández-Lafuente, R.; Jaffrezic, N.; Kuncová, G.; Matějec, V.; Scully, P. Hybrid Coatings as Transducers in Optical Biosensors. *J. Coat. Technol. Res.* **2008**, *5*, 491–496. [CrossRef]
28. Scully, P.J.; Betancor, L.; Bolyo, J.; Dzyadevych, S.; Guisan, J.M.; Fernandez-Lafuente, R.; Jaffrezic-Renault, N.; Kuncova, G.; Matějec, V.; O'Kennedy, B.; et al. Optical Fibre Biosensors Using Enzymatic Transducers to Monitor Glucose. *Meas. Sci. Technol.* **2007**, *18*, 3177. [CrossRef]
29. Koštejnová, L.; Přibyl, M.; Koštejn, M.; Kuncová, G. Analytical Characteristics of a Glucose Sensor with an Optical Oxygen Transducer for the Monitoring of Biotechnological Processes. *Meas. Sci. Technol.* **2019**, *30*, 015103. [CrossRef]
30. Tric, M.; Lederle, M.; Neuner, L.; Dolgowjasow, I.; Wiedemann, P.; Wölfl, S.; Werner, T. Optical Biosensor Optimized for Continuous In-Line Glucose Monitoring in Animal Cell Culture. *Anal. Bioanal. Chem.* **2017**, *409*, 5711–5721. [CrossRef]

Article

Hollow-Core Photonic Crystal Fiber Mach–Zehnder Interferometer for Gas Sensing †

Kaveh Nazeri [1], Farid Ahmed [2,*], Vahid Ahsani [1], Hang-Eun Joe [3], Colin Bradley [1], Ehsan Toyserkani [2] and Martin B. G. Jun [3]

1 Department of Mechanical Engineering, University of Victoria, Victoria, BC V8W 2Y2, Canada; nazerik@uvic.ca (K.N.); ahsaniv@uvic.ca (V.A.); cbr@uvic.ca (C.B.)

2 Department of Mechanical and Mechatronics Engineering, University of Waterloo, Waterloo, ON N2L 3G1, Canada; ehsan.toyserkani@uwaterloo.ca

3 School of Mechanical Engineering, Purdue University, West Lafayette, IN 47907, USA; hjoe@purdue.edu (H.-E.J.); mbgjun@purdue.edu (M.B.G.J.)

* Correspondence: farid.ahmed@uwaterloo.ca

† This paper is an extended version of our paper published in: Nazeri, K.; Ahsani, V.; Ahmed, F.; Joe, H.E.; Jun, M.; Bradley, C. Experimental comparison of the effect of the structure on MZI fiber gas sensor performance. In Proceedings of the IEEE Pacific Rim Conference on Communications, Computers and Signal Processing (PACRIM), Victoria, BC, Canada, 21–23 August 2019.

Received: 20 April 2020; Accepted: 13 May 2020; Published: 15 May 2020

Abstract: A novel and compact interferometric refractive index (RI) point sensor is developed using hollow-core photonic crystal fiber (HC-PCF) and experimentally demonstrated for high sensitivity detection and measurement of pure gases. To construct the device, the sensing element fiber (HC-PCF) was placed between two single-mode fibers with airgaps at each side. Great measurement repeatability was shown in the cyclic test for the detection of various gases. The RI sensitivity of 4629 nm/RIU was demonstrated in the RI range of 1.0000347–1.000436 for the sensor with an HC-PCF length of 3.3 mm. The sensitivity of the proposed Mach–Zehnder interferometer (MZI) sensor increases when the length of the sensing element decreases. It is shown that response and recovery times of the proposed sensor inversely change with the length of HC-PCF. Besides, spatial frequency analysis for a wide range of air-gaps revealed information on the number and power distribution of modes. It is shown that the power is mainly carried by two dominant modes in the proposed structure. The proposed sensors have the potential to improve current technology's ability to detect and quantify pure gases.

Keywords: refractive index sensor; gas sensor; hollow-core photonic crystal fiber; Mach–Zehnder interferometer

1. Introduction

Gas sensing is essential for safety and maintenance operations in many industries, including the power generation [1], petrochemical [2], and food-processing sectors [3]. For detecting the presence of gases, especially in extreme conditions, the silica optical fiber provides a promising platform, due to its unique properties. These include immunity to electromagnetic radiation [4], high-temperature durability [5], compactness, as well as high accuracy and sensitivity [6]. Researchers have pursued the applicability of optical fiber sensors across many sensing applications, because of their multifunctional sensing capabilities (e.g., refractive index (RI), temperature, and pressure) [7]. The various mechanisms that have been investigated for gas-sensing functionality include Raman scattering [8], surface Plasmon resonance [9], evanescent-field absorption [10], derivative spectroscopy [11], and interferometric sensors [6]. Successes in these research projects relied upon experimentation with a range of optical fibers: D-shaped fiber, multimode fiber, fused silica fiber optic bundles, and photonic crystal fiber (PCF) [6,8–11]. Various types of fiber optic interferometers have been studied for their RI-sensing capabilities: the Sagnac, Michelson, Fabry–Perot, and Mach–Zehnder interferometers (MZIs) [6]. Wang et al. [12] developed a micro Fabry–Perot cavity interferometer and achieved the RI sensitivity of 851 nm/RIU while having a very low-temperature sensitivity of 0.27 pm/°C and low-temperature cross-sensitivity of 3.2 E^{-7} RIU/°C. Hu et al. [13] proposed an intrinsic Fabry–Perot interferometer based on simplified hollow-core fiber and achieved a RI measurement resolution of 6.5 E^{-5}. These types of sensors typically show low insertion loss and they are relatively easy to fabricate. A Michelson interferometer was constructed by splicing a stub of large-mode-area PCF to single-mode fiber (SMF) and an RI resolution of E^{-4} in the RI range of 1.33–1.45 was reported [14]. Facile fabrication procedure and high stability over time were reported as key advantages. Sun et al. [15] proposed a hybrid interferometer by forming a Fabry–Perot cavity in one of the optical paths of the Michelson interferometer. The spectral response of this hybrid sensor allows multiparameter sensing as it has two distinct interference fringes. The simultaneous measurement capability was reported with an RI measurement resolution of 8.7 E^{-4} in the RI range of 1.33–1.38 with a temperature sensitivity of 13 pm/°C. A photonic crystal fiber Sagnac interferometer was developed by Liu et al. [16] as an RI sensor, by filling the central hole of the fiber with microfluidic analytes. Fabrication of these sensor types are complicated as filling air holes of a PCF is challenging. A high sensitivity of about 19,000 nm/RIU with a resolution of 1.05 E^{-6} was achieved in their work. MZI based optical sensors have received significant attention because they are robust, compact [17], and low-cost units that also have high levels of precision [18].

Researchers have proposed disparate configurations in fabricating in-line MZI sensors for sensing ambient RI changes. Implementation techniques already tested extend from core mismatch splicing of optical fibers [19] to cladding collapse of PCF [20], tapering of fibers [21], the use of microfiber [22], and splicing of hollow-core fiber [23]. Similarly, many approaches have been used in attempts to enhance ambient refractive index sensitivity of fiber-optic MZIs. Huang et al. [18] developed a thin-core fiber-based MZI for ammonia sensing with a sensitivity of 850 nm/RIU in the RI range of 1.5–1.518. In other studies, graphene-coated fiber-optic MZI sensors were found to have gas sensing sensitivity in the range of 3–6 pm/ppm [24,25]. Duan et al. [26] engineered a compact MZI by creating a short-length (62.5 μm) of cavity through offset-splicing the SMFs on both ends. Their innovative design resulted in a sensitivity of 3400 nm/RIU in the RI-range of 1.0 to 1.0022. PCF has also proven to be an excellent choice for fabricating RI sensors because the effective RI of the propagating cladding mode is highly sensitive to the surrounding environment [27–29]. Yang et al. [29] demonstrated the viability of a compact PCF Mach–Zehnder refractometer for sensing methane. They coated a polymer (fluoro-siloxane) over the internal surface of air holes, with one end of the PCF fusion spliced to an SMF while the other end was open for gas-molecule penetration. Through this fabrication technique, a sensitivity (defined as wavelength change per percentage of methane) of 0.514 nm%–1 was achieved [29]. This otherwise promising sensor type has drawbacks; it requires a long response time when retrieving initial conditions and also has a low level of gas selectivity.

The article by Cregan et al. in 1999 was the first research that utilized HC-PCF for the application of gas detection [30]. The presence of hollow channels in a fiber's core and cladding regions makes it difficult to fusion splice an HC-PCF to an SMF. The air holes in HC-PCF hold a large volume of air. During fusion splicing, air will expand and distort the fiber structure. In 2011, Qu et al. [31] suggested using hollow-core fiber to infiltrate various aqueous analytes in high RI measurements with a sensitivity of 1400 nm/RIU. Subsequent to this innovative proposal, a 5.1 m HC-PCF gas cell was used for the detection of methane [32]. Generally, it takes time for gas molecules to fill the cavities of HC-PCF, so this technique makes a delay in the initial measurement response to the presence of the gas [33]. Furthermore, Wynne et al.'s [34] suggestion regarding the pressure-driven filling of air-holes with gases is not applicable for real-time monitoring. Moreover, focused ion beam or femtosecond laser-assisted micro-channels can be fabricated on the cladding of HC-PCF to accelerate gas diffusion [35,36]. Nicholas et al. [37] proposed an HC-PCF-based MZI using ceramic ferrules to connect a 344-mm-long HC-PCF to two SMFs. An alternative HC-PCF-based MZI gas sensor has been reported, which employs the HC-PCF as one of the interferometer's arms [38]. Many of the sensors proposed to date either have complex configurations or poor sensitivity and response time for high-resolution measurement of gases. Ahmed et al. [39] reported a highly sensitive MZI structure that uses a small stub of HC-PCF for monitoring of CO_2; however, a detailed study on such a configuration is necessary to better understand its performances and to explore other potential applications. Recently, we studied length-dependent performance of these devices to understand their sensing properties [40]. However, more studies are required to better understand design parameters and sensing performance of these MZI sensors.

An in-line fiber optic MZI sensor, which is compact and robust with high sensitivity, is presented in this report. The HC-PCF MZI sensor utilized a short length of HC-PCF placed in between two SMFs, with gaps at each interface. The light propagation, working principles, and essential performance parameters of the proposed gas sensor are presented in this study. These include response and recovery times, RI sensitivity, as well as the number and power distribution of modes. Relative RI detection was used in all experiments, because of the difficulties in absolute RI measurement with high accuracy [41]. Experiments show promising results in the sensor's RI sensitivity. The device responds well to different gases and shows good repeatability on gas detection.

2. Working Principles

Figure 1a schematically shows a fiber arrangement of the proposed MZI sensor. A short length of HC-PCF was positioned on the V-groove and aligned with SMFs. There is an air gap at each end of the sensing element fiber. The schematic illustration of light transmission in the sensor is shown in Figure 1b. The lead-in SMF carries the incoming light wave. It radiates from the SMF core after reaching the first sensor gap in region 2 and acts as a pseudo-point light source. In the first air gap, the fundamental mode broadens and when it reaches the HC-PCF both fundamental and higher-order modes are excited in the circular channels of the sensing element. Interaction between the light and the gas molecules takes place in region 3 along the length of the sensor. Optical interference occurs in region 4 (second gap) due to the phase difference between the fundamental mode and higher-order modes. The lead-out SMF then transfers the interference spectrum to an interrogator (or spectrum analyzer). The device's reference and sensing arms are both in contact with gas molecules; however, the effect of RI change on the interference in the sensing arm is higher than in the reference arm. That imbalance occurs due to differences in optical-path lengths and phase shifts between the arms.

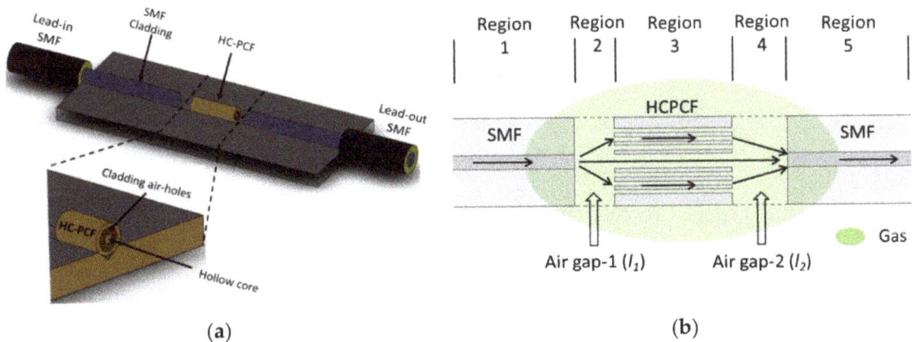

(a) (b)

Figure 1. *Cont.*

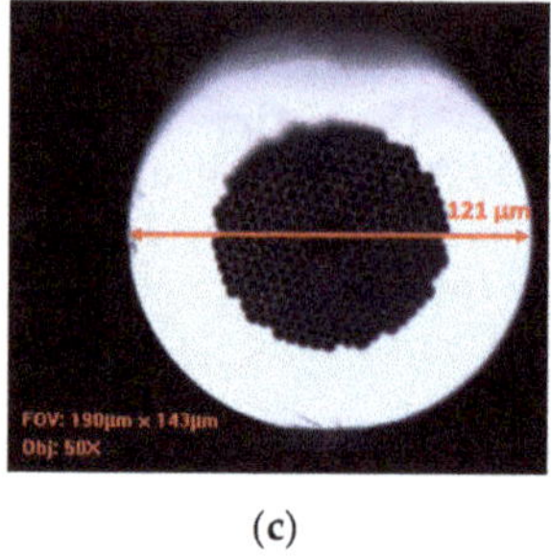

(c)

Figure 1. (**a**) Schematic of the proposed sensor arrangement, (**b**) schematic of light transmission within the sensor, and (**c**) microscopic image of the cross-section of 10-micron hollow-core photonic crystal fiber (HC-PCF) fiber. SMF = single-mode fiber.

Figure 1c shows the cross-section of the HC-PCF used in this study. This fiber offers low index guiding of light as the core-index of the HC-PCF is lower than the effective index of the cladding [42]. The photonic bandgap effect makes propagation impossible in the microstructure cladding leading to light confinement in the core. This design enhances gas sensing capabilities as the HC-PCF provides a remarkably strong interaction between gas molecules and light particles, due to strong field confinement [43,44]. Higher-order core modes and surface modes are supported by HC-PCF fibers [45]. The optical path difference between the reference arm and sensing arm defines the fiber-optic MZI sensor's interference spectrum. Such interference is a function of core intensity (I core), cladding intensity (I cladding), and phase difference (ϕ) [17,46], which can be written by the following equation:

$$I = I_{core} + I_{cladding} + 2\sqrt{I_{core} I_{cladding}} \cos \varnothing \quad (1)$$

Modes that are traveling the same distance (L) will have the phase difference (Δϕ) of:

$$\Delta\phi = 2\pi(\Delta n_{eff}) L\, \lambda^{-1} \quad (2)$$

Δn_{eff} is the difference in the effective RI between the core and cladding modes in equation 2, λ is the input wavelength, and L is the length of the HC-PCF path. Maximum transmission occurs at $\Delta\Phi = 2\pi m$ (m is an integer) and peaks forms on the transmission signal at the following wavelengths:

$$\lambda_m = (\Delta n_{eff}) L\, m^{-1} \quad (3)$$

Therefore, the mth order spectral shift can be written as:

$$\Delta\lambda_m = (\Delta n_{eff} + \Delta n) L\, m^{-1} - \Delta n_{eff}\, m^{-1} = \Delta n L\, m^{-1} \quad (4)$$

L is constant in the above equation and consequently, a change in the refractive index of the MZI's core and cladding will change Δn and correspondingly $\Delta\lambda_m$. Consequently, a shift occurs at the transmission spectrum of the device and such change can be used for sensing of a measurand.

3. Experimental Procedures

3.1. Fabrication of the MZI Sensor

Two types of fibers were used to fabricate the HC-PCF MZI sensors: the SMF (Corning SMF28) and the HC-PCF (NKT Photonics HC-PCF 1550). Lead-in and lead-out fibers are standard single-mode fibers (SMF-28) with a core diameter of 8.2 µm, numerical aperture of 0.13, and a mode field diameter (MFD) of 9.3 µm (±0.5 µm). This sensor type utilizes an NKT Photonics HC-PCF fiber (HC-PCF 1550) as the sensing element. The HC-PCF fiber has a numerical aperture (NA) of 0.2, MFD of 9.00 µm (±1 µm) and core diameter of 10.00 µm. This sensing fiber element also has cladding air holes of diameter 3.10 µm and a cladding pitch of 3.80 µm. These fibers can guide several modes within a transmission of 1490 to 1680 nm [30]. In constructing the sensor, the SMFs and HC-PCF were assembled on a standard microscope glass slide (25 mm × 5 mm × 1 mm). Micro-machining created a V-groove on the microscope glass (25 mm length, 95 µm width, and 48 µm depth) using a femtosecond laser, which is used to align fibers. A CT-30 Fujikura cleaver was used to cleave fibers. To be able to cleave short lengths of HC-PCF in the order of a few millimeters, it was necessary to extend the length of the adapter plate to decrease the distance between the cutting blade and the adapter plate. Therefore, a 4 mm long aluminum plate was machined and marks at increments of 1 mm on it. Attaching the extension plate to the adapter plate made it possible to cleave fibers with lengths down to 2 mm. The cleaved stub of HC-PCF was positioned in the middle of the V-groove and fixed using epoxy glue. The exact length of the fiber, as well as the cleaving angles on both sides of the cleaved HC-PCF, were checked by examining them under an optical tooling microscope. Afterward, the single-mode fibers were positioned in fiber holders mounted on linear-translation micro stages and aligned with the sensing element fiber on the V-groove. Figure 2a shows an isometric view of the fabrication setup. To achieve a strong interference spectrum, gap lengths on both sides of HC-PCF were accurately adjusted. In this way mode splitting and recombination can be controlled. Fibers were then glued to microscope glass when an acceptable signal was observed. To provide mechanical strength to the assembly, the glass slide was secured in a meshed stainless steel tube, as shown in Figure 2b. Testing proved the robust effectiveness of the resulting sensor. Spacing between the HC-PCF and SMFs enabled ambient gas to diffuse into the HC-PCF air holes.

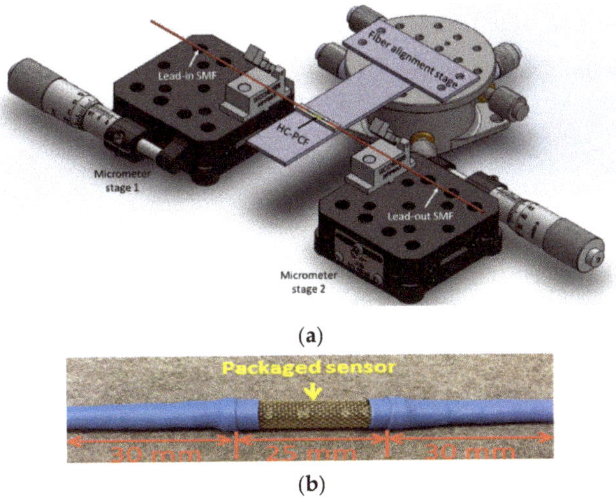

Figure 2. (**a**) Isometric view of the fabrication setup using two linear-translation micro stages for accurate control of gap distances, (**b**) packaged sensor using meshed stainless steel tube.

The normalized transmission spectrum of a sensor with an HC-PCF length of 3.30 mm and a gap distance of 1 mm on each side (Sensor C) is shown in Figure 3a. Figure 3b shows the fringe spacing of the same sensor. The measurement was taken when the device was immersed in Nitrogen (99.99% pure, atmospheric pressure) at room temperature. Each valley measured at the sensor's output, see Figure 3a, results from interference between the signal arms in the MZI at that wavelength. The magnified spectrum graph shows a fringe spacing of 1.91 nm and a full width at half maximum (for transmission dip) of 0.47 nm. For the same configurations, the fringe spacings of sensor A (L = 4.97 mm) and sensor

B (L = 4.73 mm) are 1.70 nm and 1.74 nm, respectively. The fringe spacing of the transmission spectrum increases as the length of HC-PCF decreases.

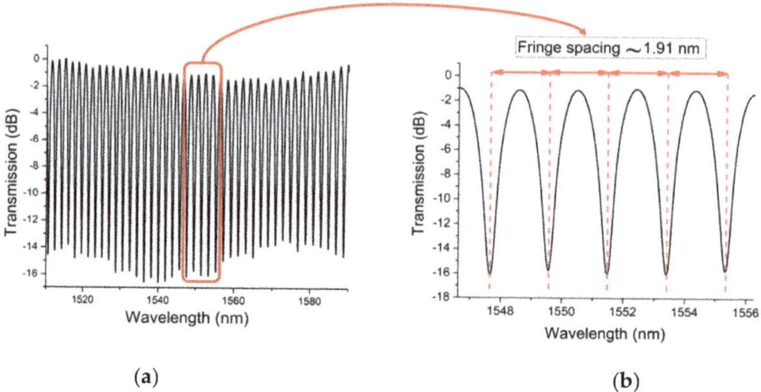

Figure 3. (**a**) Normalized transmission spectrum of an HC-PCF Mach–Zehnder interferometer (MZI) sensor with HC-PCF length of 3.3 mm and gaps of 1mm immersed in Nitrogen at room temperature and atmospheric pressure, (**b**) fringe spacing of the same sensor.

3.2. Spatial Frequency Analysis

In order to analyze the modes participating in the modal interference process, the transmission spectrum of MZIs with 4 mm of HC-PCF as a sensing element was Fourier transformed. This process allowed us to obtain the sensor's corresponding spatial frequency, described as $v = \frac{\Delta n_{eff} \cdot D}{\lambda^2}$ [47], where Δn_{eff} represents the effective RI-difference between core and cladding of the sensing element and D is the distance between SMFs at each of the sensor's ends. D varies from 4 mm to 16 mm in 500-micron increments. Different peaks in the spatial frequency graph correspond to the interference between the fundamental mode and different higher-order modes.

Testing the MZIs with 10 µm HC-PCF as their sensing element revealed several multimodal-interference patterns occurring in the transmission spectrum. Further, in such a sensor, power is mainly distributed between two dominant modes in the spatial frequency spectrum, a finding that holds true across the entire range of gap distances. This phenomenon confirms that higher-order modes would gradually leak off the sensing fiber, contributing to transmission losses. So, fewer peaks would turn up in the spatial frequency graph due to a weakening interference-effect. As an example of the described effect, Figure 4a presents the spatial frequency graph for an MZI with 4 mm of 10 µm HC-PCF and gaps of 1.5 mm on each side (D = 7 mm). The sensor has a strong cladding mode with a spatial frequency of 5×10^{-4} (1/nm) and a normalized fast Fourier transform (FFT) value of 3.14: labeled core-cladding 1. Besides this dominant cladding mode, the sensor has a relatively weaker cladding mode (core-cladding 2) with a spatial frequency of 1.1×10^{-3} (1/nm) and a normalized FFT value of 0.99. Experimental findings show that for gaps from 0 to 1.65 mm, core-cladding 1 is the dominant cladding mode, while for higher gaps core-cladding 2 became the dominant mode. The highest power transmission resulted in MZIs with gaps of 1.35 mm, and the amplitude of spatial frequencies was seen to decrease intensely for gaps greater than 4.5 mm. Figure 4b was plotted by tracking dominant modes to show how the magnitude of spatial frequencies increases by increasing gap lengths for this sensor.

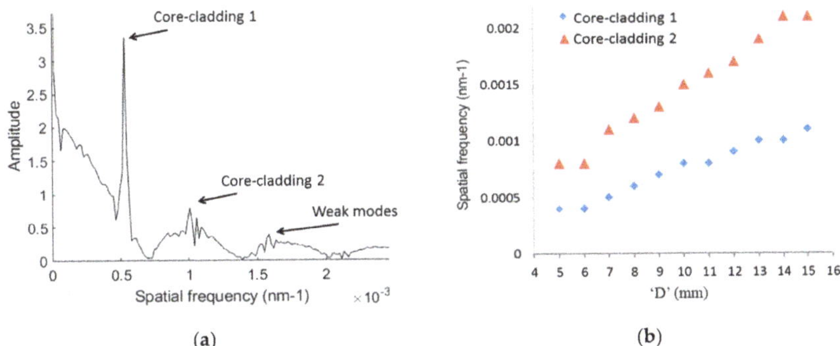

Figure 4. (**a**) Spatial frequency graph for MZI with HC-PCF length of 4 mm and D of 7 mm, (**b**) tracking dominant modes of the sensor for D. D = length of the sensing element.

3.3. Characterization

In the first set of experiments, RI measurements using three MZIs (constructed with different lengths of HC-PCF as their sensing elements) were carried out and their relative performances were compared. Figure 5 schematically shows the sensor evaluation system that includes the optical interrogator, a circulator, the MZI sensor, a Fiber Bragg Grating (FBG), reference gas tank, and measurand gas tanks. The MZI sensors under investigation were placed in a chamber with four gas intake valves. Reference nitrogen (N_2), and measurand gas-tanks ('He', 'Ar', and 'CH_4') were connected to these valves. The experiment used helium, methane, and argon with purity levels of 99.999%, 99%, and 99.99%, respectively. Using pressure regulators, an injection pressure of 15 psi was maintained during the testing process. To maintain constant pressure in the test chamber a discharge tube with a bubbler was connected to the test chamber. An interrogator (SM125) with a resolution of 1 pm was used to record and evaluate changes in the transmission spectrum. In addition, a FBG (sensitivity ~10 pm/°C) was positioned in the chamber to monitor and record the temperature variations. The spectral shifts of three sensor types and FBG were analyzed using the Micron Optics' Enlight software. The experiments started with injecting N_2 into the test chamber for long enough time to make sure an even gas diffusion into the air holes of HC-PCF was achieved. Measurand gases were then injected into the chamber ('He', 'Ar', or 'CH_4'). Using the mentioned software, spectral responses were recorded. Response and recovery times as well as refractive index sensitivity are among important sensing performance parameters of a gas sensor and were studied for three MZIs. The cyclic tests were performed using the various sensors to inspect the repeatability of RI measurements. Temperature, pressure and the injected gas species determined the spectral response of each sensor. Therefore, MZI sensors were temperature-characterized to compensate for the effect of temperature fluctuations during the experiments.

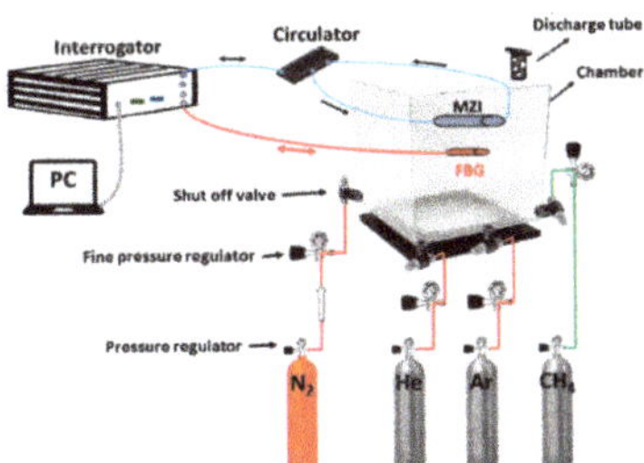

Figure 5. Schematic of the experimental setup; tests were carried out at atmospheric pressure and room temperature.

Another set of experiments sought to analyze the effect of gap distances on modal interference in the proposed MZI gas sensor. Here, lead-in and lead-out SMFs were not glued to the glass to facilitate easy adjustment of both airgaps. Using linear micro stages, gap lengths increased from 0 to 6 mm in 500-micron increments. Ensuring equal gap distance on both sides, we collected transmission spectrums for an interferometer with 10 μm HC-PCF as its sensing fiber. Spectrums were Fourier transformed to produce spatial frequency graphs, to explore the power distribution and the number of the sensor's modes.

4. Results and Discussion

4.1. Refractive Index Sensing

Figure 6a illustrates the responses of sensor A (L = 4.97 mm) to methane, argon, and helium for one cycle. MZI sensors were exposed to measurand gases separately, to determine its spectral response to each gas. The sensor was interrogated with each measurand gas to investigate its spectral response in a complete test cycle. Each cycle started with the injection of Nitrogen (99.99% pure) until saturation followed by injection and measurement of target gas; and finally, an injection of Nitrogen back into the chamber, to purge the gas. The injection of gases was carried out for 7 minutes at each stage of a test cycle. As shown in Figure 6, the ambient gas in the test chamber determines the sensor's wavelength response. Considering the location of the spectrum in N_2 as the reference, sensor A showed spectrum shifts of 780 pm (±6 pm) when immersed in helium, 45 pm (±1 pm) when immersed in argon, and 440 pm (± 3 pm) when immersed in CH_4. Spectral shifts of three valleys at different wavelengths were used to estimate mean wavelength shifts and measurement errors. Redshifts were recorded in the transmission spectrum for Ar or He and blue shifts were recorded for CH_4. This finding can be explained in terms of spectral response to RI change. For a given ambient RI, the sensor's transmission spectrum shows redshift to a negative RI change and blue shift to a positive RI change. In standard conditions, the RI values of He, Ar, N_2, and CH_4 are 1.0000347, 1.0002820, 1.0002944, and 1.0004365, respectively. The interference fringe showed a redshift in the presence of helium and argon because RI of nitrogen is higher than their RIs. In contrast, the spectrum underwent a blue shift for methane, as the RI of nitrogen is lower than the RI of methane. The transmission fringe shifts of MZI sensors for helium, methane, and argon are listed in Table 1.

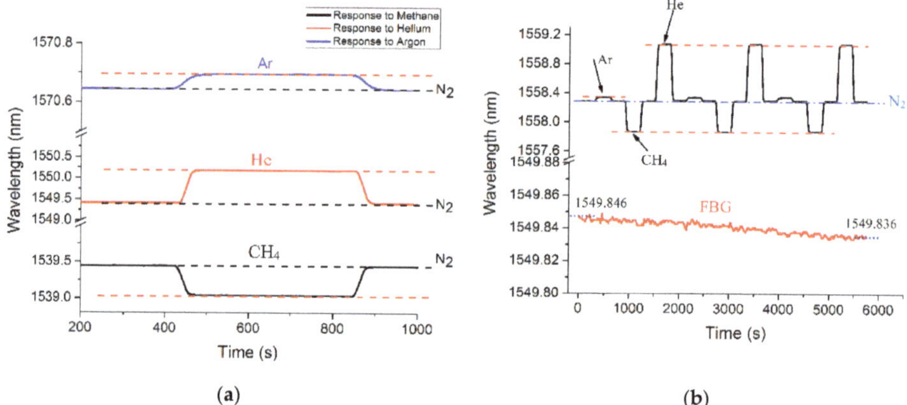

Figure 6. (**a**) The spectral shifts of sensor A when immersed in argon, helium, and methane injections, (**b**) the results of sequential sensing of measurand gases with sensor A, with gas injections carried out in the sequence of argon, methane, and helium.

Table 1. Transmission fringe shift of the MZI sensors for helium, methane, and argon.

Sensor	HC-PCF Length	Spectral Shift (pm) in Helium	Spectral Shift (pm) in Methane	Spectral Shift (pm) in Argon	RI Sensitivity (nm/RIU)
A	4.97 mm	780	440 (negative)	45	3019
B	4.73 mm	1060	600 (negative)	70	4300
C	3.30 mm	1300	618 (negative)	100	4629

Sensor C, which has the shortest length of HC-PCF, shows the highest wavelength shifts among the three sensors tested when interrogated with all three gases. In contrast, sensor A, which has the longest HC-PCF stub of the three sensors, shows the smallest shifts. The RI sensitivities of the interferometric sensors are listed in Table 1, all falling in the RI range of 1.0000347–1.0004365. This RI range was selected based on the availability of gas tanks, and it could be extended in future research. The highest sensitivity was achieved by sensor C: 4629 (nm/RIU). This suggests that the RI sensitivity of the HC-PCF MZI sensors increases as the length of the HC-PCF stub decreases. As the next step in our experiments, argon, methane, and helium gases were sequentially injected into the test chamber, to investigate the sensors' capacities to detect multiple gases. In each test cycle, the gas injection was carried out in the sequence of N_2, Ar, N_2, CH_4, N_2, He, and N_2. This sequence was then repeated three times to determine sensing repeatability. Figure 6b shows the sequential gas response for sensor A, where the test cycles produced identical results. An FBG was used to record any temperature variation during the test. A maximum temperature fluctuation of 1 °C was recorded during the entire experiment.

To check the consistency of the sensor's measurements, repeatability tests were performed using all three sensors. For each test cycle, the sequential injection of nitrogen, measurand gas, and nitrogen was performed at 5 minutes intervals. Figure 7a shows the repeatability of sensing helium gas using all three sensors for eight cycles. The repeatability test for sensing methane gas was conducted for three test-cycles, as shown in Figure 7b. Both graphs below show the normalized wavelength shift that resulted when the chamber was sequentially filled with nitrogen and measurand gases. The data shows great repeatability of gas detection using the proposed HC-PCF interferometer.

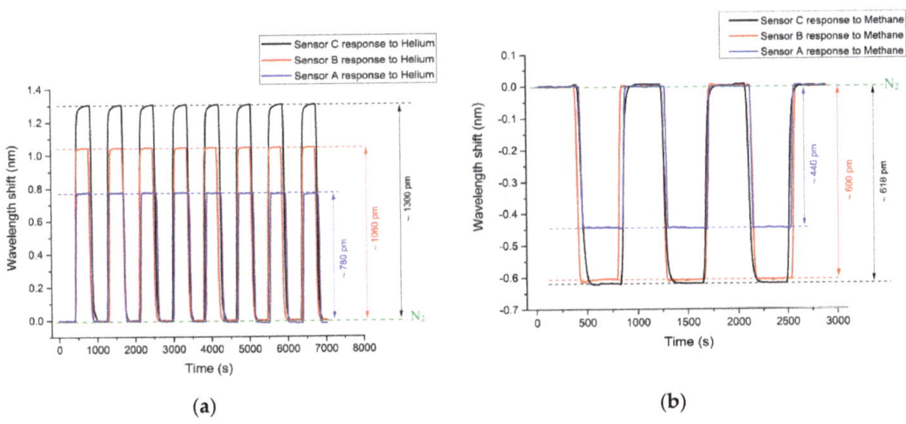

Figure 7. The normalized cyclic response of HC-PCF MZI sensors to (**a**) helium and (**b**) methane.

The RI sensitivity of sensors A, B, and C are 3019 nm/RIU, 4300 nm/RIU, and 4629 nm/RIU, respectively. Figure 8 shows the RI sensitivity of sensor A in the mentioned RI range. These data points were obtained via five separate measurements with a measurement error of $\pm 1\ E^{-6}$, $\pm 2.3\ E^{-6}$, and $\pm 5\ E^{-7}$ for methane, helium, and argon, respectively. The proposed sensor configuration can improve on current technology, due to its linear RI response and high sensitivity to gases. The proposed interferometric sensor has, nonetheless, the potential for advancing current capacity for gas detection, quantitatively analyzing changes in pure gases as well as environmental monitoring applications. The RI characterization tests were conducted using an optical interrogator that has a wavelength accuracy of 1 pm (0.001 nm). Therefore, the sensor C (sensitivity of 4629 nm/RIU) has a RI resolution of 2.1 E^{-7}. Similarly, the sensing resolution of sensors A and B can be calculated.

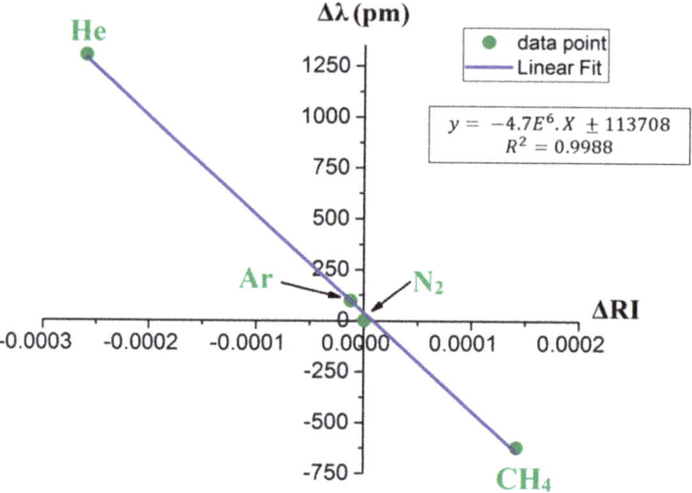

Figure 8. Sensitivity graph for sensor C to ambient RI change.

The refractive index of any target gas (RI $_{\text{target gas}}$) can be written as:

$$\text{RI}_{\text{target gas}} = \text{RI}_{\text{N}_2} - \Delta\text{RI} = \text{RI}_{\text{N}_2} - \Delta\lambda/(\text{RI sensitivity}) \quad (5)$$

The spectral shift, Δλ in the above equation can be attained by tracking valleys of transmission fringe of a sensor, as shown in Figures 6 and 7. RI_{N_2} is the refractive index of nitrogen, and ΔRI is the relative difference in RI between nitrogen and measurand gas. By knowing the wavelength shift (Δλ) and sensitivity of the MZI sensor, ΔRI can be calculated.

Table 2 compares the sensitivity achieved in the present research with other similar and alternative studies available in published works. The table shows that the proposed MZI configuration shows much higher sensitivity in gas sensing compared to its counterparts in the RI range of 1 to 1.02. As shown in [40], decreasing the length of HC-PCF, the sensitivity of this sensor can be further improved. The proposed sensor is fairly compact (3.3 mm) compared to other HC-PCF based RI sensors [31,32,37,38], some of which are as long as ~35 cm. Therefore, the proposed MZI configuration is believed to perform much better in single-point gas sensing. It is worth mentioning here that even though a compact Fabry–Perot fiber sensor (in the range of micrometer) can be fabricated using ultrafast laser micromachining they have relatively poor RI sensitivity [12]. Despite its excellent gas sensing capabilities, the reported device has few drawbacks including fabrication complexity as it requires alignment and positing of the HC-PCF stub and cross-sensitivity to other measurands such as temperature and pressure. With the recent improvement in automated fiber alignment and positing systems, we believe the fabrication complexity can be drastically reduced for commercial applications. Similar to other fiber-optic sensors, the cross-sensitivities can be eliminated or reduced using an in-line fiber sensor such as a properly packaged FBG. The demonstrated sensor also needs to be packaged with a suitable membrane for selective sensing of gasses.

Table 2. RI sensitivity comparison for gas sensing with other reported fiber-optic gas sensors.

Optical Structure	RI Range	RI Sensitivity (nm/RIU)	Reference
Proposed HC-PCF MZI	1.000034–1.000449	4629	This work
HC-PCF MZI	1.0000–1.0005	1233	[37]
Fabry-Perot (FP) based on hollow silica tube	1.00027–1.00189 1.00007–1.00051	1546	[48]
Surface plasmon resonance (SPR) with metallic surface grating (tapered SMF)	1–1.41	500	[49]
Hybrid optical fiber FP interferometer	1.0005–1.00275	560	[50]
SPR based on fiber grating in multi-mode fiber	1–1.33	280	[51]
Cavity based FP	1.0000–1.0025	1053	[52]
Open cavity MZI	1–1.02	3402	[26]

4.2. Sensor Response and Recovery Times

Figure 9 illustrates the response and recovery times of sensor A for one cycle of methane sensing. The time duration that an MZI device takes to reach 90% of the total wavelength shifts is defined as response/recovery times. Accordingly, response and recovery times of sensor A are 32 s and 39 s for methane. Response and recovery times of three HC-PCF MZI sensors to methane, helium, and argon are listed in Table 3. Each reported time in this table is an average of five response or recovery times. Results indicate that sensor A, which has the longest HC-PCF stub, shows the fastest response/recovery times. However, the highest RI sensitivity was achieved using sensor C, which has the shortest length of HC-PCF. Response and recovery times depend on HC-PCF lengths and the volume of the test chamber. The test chamber has a dimension of 14.5 cm × 11.2 cm × 4.4 cm.

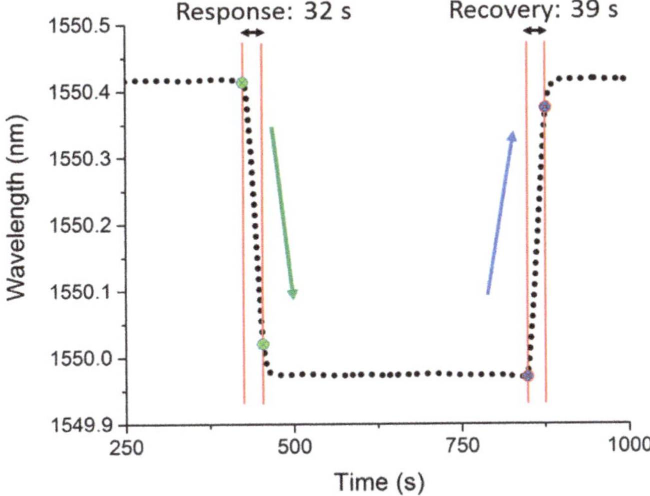

Figure 9. Response and recovery times of sensor A for methane.

Table 3. Response and recovery times of HC-PCF MZI sensors to different gases.

HC-PCF Length (mm)	A (4.97)	B (4.73)	C (3.30)
Helium: response (s)/recovery (s)	50/50	50/55	57/57
Methane: response (s)/recovery (s)	32/39	44/46	46/56
Argon: response (s)/recovery (s)	37/44	62/49	110/100

4.3. Temperature Characterization

The RI of a gas depends not only on gas species but also on ambient temperature and pressure. All the experiments were conducted at atmospheric pressure and room temperature. However, fluctuation of ~1 °C was recorded using an FBG sensor during the experiments, a result shown in Figure 6b. Therefore, it is required to characterize the temperature sensitivity of the HC-PCF MZI sensor before deploying the sensor for applications in the field. As part of the present research, HC-PCF sensors were placed in an oven, and the temperature was varied from 35 °C to 65 °C in 10 °C increments. Figure 10 displays the resulting correlation between recorded wavelength shifts and measured temperatures of the sensors and FBG. The temperature sensitivities of sensors A, B, and C were found to be 33.1 pm/°C, 31.6 pm/°C, and 20 pm/°C, respectively. This finding shows that the temperature sensitivity of the fiber-optic interferometer decreases when the length of the HC-PCF decreases. As shown in Figure 10, a typical FBG has a temperature sensitivity of 10 pm/°C and it is insensitive to ambient RI change. Therefore, an in-line or parallel FBG can be placed as a reference to eliminate temperature cross-sensitivity in ambient RI measurement for practical applications. Like temperature, a fiber-optic pressure gauge that is insensitive to ambient RI can be used to eliminate pressure cross-sensitivity in real-life measurement.

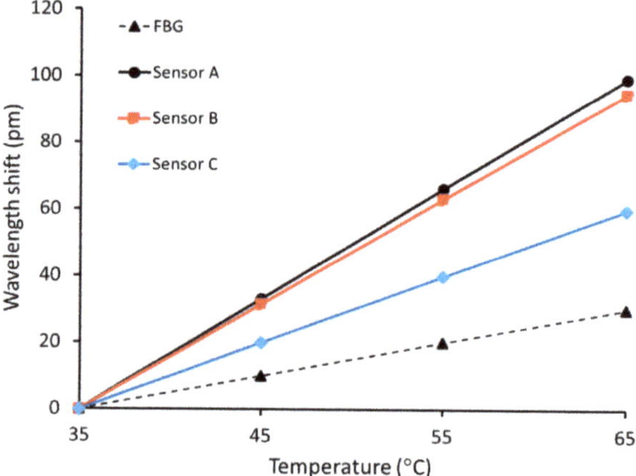

Figure 10. Temperature characterization of Fiber Bragg Grating (FBG) and HC-PCF sensors.

5. Conclusions

A compact fiber-optic MZI sensor is proposed and has been experimentally demonstrated for ultra-high sensitive detection of gases. Different lengths of HC-PCF stubs were used to construct and characterize several sensors. The resulting MZI sensors were able to measure the RI of target gases and showed great sensitivity to measurand gases. The Refractive index sensitivity of 4629 nm/RIU was achieved for the MZI with an HC-PCF length of 3.30 mm. The RI sensitivity of the proposed MZI sensor inversely relates to the length of the HC-PCF stub. However, response and recovery times turned out to be shorter for longer HC-PCF stubs. The effect of gap distances on the number and amplitude distribution of the sensors' modes was examined, and spatial frequency analysis revealed that power is mainly carried by two dominant modes in the proposed MZI. These novel and compact sensors have high-temperature sensitivity, compared to an FBG. With appropriate packaging, the proposed sensor becomes robust and is a suitable choice for low-percentage detection of gases as well as environmental monitoring.

Author Contributions: K.N. and F.A. proposed and designed the set of experiments; K.N., V.A., and F.A. performed the experiments; K.N., F.A., H.-E.J., and V.A. analyzed the experimental results, K.N. wrote the original draft of the manuscript; K.N., F.A., C.B., M.B.G.J., and E.T. reviewed the manuscript; M.B.G.J., E.T., and C.B. supervised the project. All authors have read and agreed to the published version of the manuscript.

Funding: This research received no external funding.

Acknowledgments: The Natural Sciences and Engineering Research Council (NSERC) of Canada and Korea Carbon Capture and Sequestration R&D Center (KCRC) supported this work.

Conflicts of Interest: The authors declare no conflict of interest.

References

1. Liu, Y.; Parisi, J.; Sun, X.; Lei, Y. Solid-state gas sensors for high temperature applications a review. *J. Mater. B Chem.* **2014**, *2*, 9919–9943. [CrossRef]
2. Moser, H.; Pölz, W.; Waclawek, J.P.; Ofner, J.; Lendl, B. Implementation of a quantum cascade laser-based gas sensor prototype for sub-ppmv H2S measurements in a petrochemical process gas stream. *Anal. Bioanal. Chem.* **2016**, *409*, 729–739. [CrossRef] [PubMed]
3. Shi, Y.; Li, Z.; Shi, J.; Zhang, F.; Zhou, X.; Li, Y.; Holmes, M.; Zhang, W.; Zou, X. Titanium dioxide-polyaniline/silk fibroin microfiber sensor for pork freshness evaluation. *Sens. Actuators B Chem.* **2018**, *260*, 465–474. [CrossRef]
4. Aoni, R.A.; Ahmed, K.; Asaduzzaman, S.; Paul, B.K.; Ahmed, R. Development of Photonic Crystal Fiber-Based Gas/Chemical Sensors. *Comput. Photonic Sens.* **2018**, 287–317. [CrossRef]
5. Pinto, A.M.R.; Lopez-Amo, M. Photonic Crystal Fibers for Sensing Applications. *J. Sens.* **2012**, *2012*, 1–21. [CrossRef]
6. Hu, D.J.J.; Wong, R.Y.-N.; Shum, P.P. Photonic Crystal Fiber–Based Interferometric Sensors. In *Selected Topics on Optical Fiber Technologies and Applications*; InTech: London, UK, 2018; pp. 22–41.
7. Joe, H.-E.; Yun, H.; Jo, S.; Jun, M.B.; Min, B.-K. A review on optical fiber sensors for environmental monitoring. *Int. J. Precis. Eng. Manuf. Technol.* **2018**, *5*, 173–191. [CrossRef]
8. Buric, M.P.; Chen, K.P.; Falk, J.; Woodruff, S.D. Enhanced spontaneous Raman scattering and gas composition analysis using a photonic crystal fiber. *Appl. Opt.* **2008**, *47*, 4255–4261. [CrossRef]
9. AbdelGhani, A.; Chovelon, J.; Jaffrezic-Renault, N.; Ronot-Trioli, C.; Veillas, C.; Gagnaire, H. Surface plasmon resonance fibre-optic sensor for gas detection. *Sens. Actuators B Chem.* **1997**, *39*, 407–410. [CrossRef]
10. Stewart, G.; Muhammad, F.; Culshaw, B. Sensitivity improvement for evanescent-wave gas sensors. *Sens. Actuators B Chem.* **1993**, *11*, 521–524. [CrossRef]
11. Yang, X.; Chang, A.S.P.; Chen, B.; Gu, C.; Bond, T.C. Multiplexed gas sensing based on Raman spectroscopy in photonic crystal fiber. In Proceedings of the IEEE Photonics Conference 2012, Institute of Electrical and Electronics Engineers (IEEE), Burlingame, CA, USA, 23–27 September 2012; pp. 447–448.
12. Wang, Y.; Wang, D.N.; Liao, C.R.; Hu, T.; Guo, J.; Wei, H. Temperature-insensitive refractive index sensing by use of micro Fabry-Pérot cavity based on simplified hollow-core photonic crystal fiber. *Opt. Lett.* **2013**, *38*, 269–271. [CrossRef] [PubMed]
13. Hu, D.J.J.; Wang, Y.; Lim, J.L.; Zhang, T.; Milenko, K.; Chen, Z.; Jiang, M.; Wang, G.; Luan, F.; Shum, P.; et al. Novel Miniaturized Fabry–Perot Refractometer Based on a Simplified Hollow-Core Fiber with a Hollow Silica Sphere Tip. *IEEE Sens. J.* **2011**, *12*, 1239–1245. [CrossRef]
14. Jha, R.; Villatoro, J.; Badenes, G. Ultrastable in reflection photonic crystal fiber modal interferometer for accurate refractive index sensing. *Appl. Phys. Lett.* **2008**, *93*, 191106. [CrossRef]
15. Sun, H.; Zhang, J.; Rong, Q.; Feng, D.; Du, Y.; Zhang, X.; Su, D.; Zhou, L.; Feng, Z.; Qiao, X.; et al. A Hybrid Fiber Interferometer for Simultaneous Refractive Index and Temperature Measurements Based on Fabry–Perot/Michelson Interference. *IEEE Sens. J.* **2013**, *13*, 2039–2044. [CrossRef]
16. Liu, Q.; Xin, L.; Wu, Z.; Xing, L. Refractive index sensor of a photonic crystal fiber Sagnac interferometer based on variable polarization states. *Appl. Phys. Express* **2019**, *12*, 062009. [CrossRef]
17. Li, L.; Xia, L.; Xie, Z.; Liu, D. All-fiber Mach-Zehnder interferometers for sensing applications. *Opt. Express* **2012**, *20*, 11109–11120. [CrossRef]
18. Huang, X.; Li, X.; Yang, J.; Tao, C.; Guo, X.; Bao, H.; Yin, Y.; Chen, H.; Zhu, Y. An in-line Mach-Zehnder Interferometer Using Thin-core Fiber for Ammonia Gas Sensing with High Sensitivity. *Sci. Rep.* **2017**, *7*, 44994. [CrossRef]
19. Choi, H.Y.; Kim, M.J.; Lee, B.-H. All-fiber Mach-Zehnder type interferometers formed in photonic crystal fiber. *Opt. Express* **2007**, *15*, 5711–5720. [CrossRef]
20. Ahmed, F.; Ahsani, V.; Melo, L.; Wild, P.; Jun, M.B.G.; Melo, L. Miniaturized Tapered Photonic Crystal Fiber Mach-Zehnder Interferometer for Enhanced Refractive Index Sensing. *IEEE Sens. J.* **2016**, *16*, 1. [CrossRef]

21. Ahsani, V.; Ahmed, F.; Jun, M.B.; Bradley, C. Tapered Fiber-Optic Mach-Zehnder Interferometer for Ultra-High Sensitivity Measurement of Refractive Index. *Sensors* **2019**, *19*, 1652. [CrossRef]
22. Ahmed, F.; Ahsani, V.; Saad, A.; Jun, M.B.G. Bragg Grating Embedded in Mach-Zehnder Interferometer for Refractive Index and Temperature Sensing. *IEEE Photon Technol. Lett.* **2016**, *28*, 1968–1971. [CrossRef]
23. Hu, L.M.; Chan, C.C.; Dong, X.; Wang, Y.P.; Zu, P.; Wong, W.C.; Qian, W.W.; Li, T. Photonic Crystal Fiber Strain Sensor Based on Modified Mach–Zehnder Interferometer. *IEEE Photon J.* **2011**, *4*, 114–118. [CrossRef]
24. Yao, B.; Wu, Y.; Cheng, Y.; Zhang, A.; Gong, Y.; Rao, Y.-J.; Wang, Z.-G.; Chen, Y. All-optical Mach–Zehnder interferometric NH3 gas sensor based on graphene/microfiber hybrid waveguide. *Sens. Actuators B Chem.* **2014**, *194*, 142–148. [CrossRef]
25. Hao, T.; Chiang, K.S. Graphene-Based Ammonia-Gas Sensor Using In-Fiber Mach-Zehnder Interferometer. *IEEE Photon Technol. Lett.* **2017**, *29*, 2035–2038. [CrossRef]
26. Duan, D.; Rao, Y.; Xu, L.-C.; Zhu, T.; Wu, D.; Yao, J. In-fiber Mach–Zehnder interferometer formed by large lateral offset fusion splicing for gases refractive index measurement with high sensitivity. *Sens. Actuators B Chem.* **2011**, *160*, 1198–1202. [CrossRef]
27. Zhang, T.; Zheng, Y.; Wang, C.; Mu, Z.; Liu, Y.; Lin, J. A review of photonic crystal fiber sensor applications for different physical quantities. *Appl. Spectrosc. Rev.* **2017**, *53*, 486–502. [CrossRef]
28. Tao, C.; Wei, H.; Feng, W.-L. Photonic crystal fiber in-line Mach-Zehnder interferometer for explosive detection. *Opt. Express* **2016**, *24*, 2806–2817. [CrossRef]
29. Yang, J.; Zhou, L.; Che, X.; Huang, J.; Li, X.; Chen, W. Photonic crystal fiber methane sensor based on modal interference with an ultraviolet curable fluoro-siloxane nano-film incorporating cryptophane A. *Sens. Actuators B Chem.* **2016**, *235*, 717–722. [CrossRef]
30. Cregan, R.F. Single-Mode Photonic Band Gap Guidance of Light in Air. *Science* **1999**, *285*, 1537–1539. [CrossRef]
31. Qu, H.; Skorobogatiy, M. Liquid-core low-refractive-index-contrast Bragg fiber sensor. *Appl. Phys. Lett.* **2011**, *98*, 201114. [CrossRef]
32. Cubillas, A.M.; Silva-Lopez, M.; Lázaro, J.M.; Conde, O.M.; Petrovich, M.N.; López-Higuera, J.M. Detection of methane at 1670-nm band with a hollow-core photonic bandgap fiber. *Photonics Eur.* **2008**, *6990*, 69900.
33. Jin, W.; Ho, H.; Cao, Y.; Ju, J.; Qi, L. Gas detection with micro- and nano-engineered optical fibers. *Opt. Fiber Technol.* **2013**, *19*, 741–759. [CrossRef]
34. Wynne, R.; Barabadi, B.; Creedon, K.J.; Ortega, A. Sub-Minute Response Time of a Hollow-Core Photonic Bandgap Fiber Gas Sensor. *J. Light. Technol.* **2009**, *27*, 1590–1596. [CrossRef]
35. Li, X.; Pawłat, J.; Liang, J.; Xu, G.; Ueda, T. Fabrication of Photonic Bandgap Fiber Gas Cell Using Focused Ion Beam Cutting. *Jpn. J. Appl. Phys.* **2009**, *48*, 06FK05. [CrossRef]
36. Wang, D.N. Micro-engineered optical fiber sensors fabricated by femtosecond laser micromachining. *Internat. Opt. Fabricat. Test.* **2012**. [CrossRef]
37. Andrews, N.L.P.; Ross, R.; Munzke, D.; Van Hoorn, C.; Brzezinski, A.; Barnes, J.A.; Reich, O.; Loock, H.-P. In-fiber Mach-Zehnder interferometer for gas refractive index measurements based on a hollow-core photonic crystal fiber. *Opt. Express* **2016**, *24*, 14086. [CrossRef]
38. Shavrin, I.; Novotny, S.; Shevchenko, A.; Ludvigsen, H. Gas refractometry using a hollow-core photonic bandgap fiber in a Mach-Zehnder-type interferometer. *Appl. Phys. Lett.* **2012**, *100*, 51106. [CrossRef]
39. Ahmed, F.; Ahsani, V.; Nazeri, K.; Marzband, E.; Bradley, C.; Toyserkani, E.; Jun, M.B.G. Monitoring of Carbon Dioxide Using Hollow-Core Photonic Crystal Fiber Mach-Zehnder Interferometer. *Sensors* **2019**, *19*, 3357. [CrossRef]
40. Nazeri, K.; Ahsani, V.; Ahmed, F.; Joe, H.-E.; Jun, M.; Bradley, C. Experimental comparison of the effect of the structure on MZI fiber gas sensor performance. In Proceedings of the 2019 IEEE Pacific Rim Conference on Communications, Computers and Signal Processing (PACRIM), Institute of Electrical and Electronics Engineers (IEEE), Victoria, BC, Canada, 21–23 August 2019.
41. Xiao, G.; Adnet, A.; Zhang, Z.; Sun, F.G.; Grover, C.P. Monitoring changes in the refractive index of gases by means of a fiber optic Fabry-Perot interferometer sensor. *Sens. Actuators A Phys.* **2005**, *118*, 177–182. [CrossRef]
42. Benabid, F.; Roberts, P. Linear and nonlinear optical properties of hollow core photonic crystal fiber. *J. Mod. Opt.* **2011**, *58*, 87–124. [CrossRef]

43. Cordeiro, C.M.D.B.; Franco, M.; Chesini, G.; Barretto, E.C.S.; Lwin, R.; Cruz, C.H.D.B.; Large, M.C.J. Microstructured-core optical fibre for evanescent sensing applications. *Opt. Express* **2006**, *14*, 13056. [CrossRef] [PubMed]
44. Zhi-Guo, Z.; Fang-Di, Z.; Min, Z.; Pei-Da, Y. Gas sensing properties of index-guided PCF with air-core. *Opt. Laser Technol.* **2008**, *40*, 167–174. [CrossRef]
45. Tao, C.; Li, X.; Yang, J.; Shi, Y. Optical fiber sensing element based on luminescence quenching of silica nanowires modified with cryptophane-A for the detection of methane. *Sens. Actuators B Chem.* **2011**, *156*, 553–558. [CrossRef]
46. Wang, Q.; Wei, W.; Guo, M.; Zhao, Y. Optimization of cascaded fiber tapered Mach–Zehnder interferometer and refractive index sensing technology. *Sens. Actuators B Chem.* **2016**, *222*, 159–165. [CrossRef]
47. Zhang, H.; Gao, S.; Luo, Y.; Chen, Z.; Xiong, S.; Wan, L.; Huang, X.; Huang, B.; Feng, Y.; He, M.; et al. Ultrasensitive Mach-Zehnder Interferometric Temperature Sensor Based on Liquid-Filled D-Shaped Fiber Cavity. *Sensors* **2018**, *18*, 1239. [CrossRef]
48. Jia, P.; Fang, G.; Liang, T.; Hong, Y.; Tan, Q.; Chen, X.; Liu, W.; Xue, C.; Liu, J.; Zhang, W.; et al. Temperature-compensated fiber-optic Fabry–Perot interferometric gas refractive-index sensor based on hollow silica tube for high-temperature application. *Sens. Actuators B Chem.* **2017**, *244*, 226–232. [CrossRef]
49. Ding, W.; Andrews, S.R.; Birks, T.A.; Maier, S.A. Modal coupling in fiber tapers decorated with metallic surface gratings. *Opt. Lett.* **2006**, *31*, 2556–2558. [CrossRef]
50. Wang, R.; Qiao, X. Hybrid optical fiber Fabry-Perot interferometer for simultaneous measurement of gas refractive index and temperature. *Appl. Opt.* **2014**, *53*, 7724–7728. [CrossRef]
51. Nemova, G.; Kashyap, R. Novel fiber Bragg grating assisted plasmon-polariton for bio-medical refractive-index sensors. *J. Mater. Sci. Mater. Electron.* **2007**, *18*, 327–330. [CrossRef]
52. Ferreira, M.S.; Coelho, L.; Schuster, K.; Kobelke, J.; Santos, J.L.; Frazão, O. Fabry–Perot cavity based on a diaphragm-free hollow-core silica tube. *Opt. Lett.* **2011**, *36*, 4029–4031. [CrossRef]

Letter

Label-Free and Reproducible Chemical Sensor Using the Vertical-Fluid-Array Induced Optical Fiber Long Period Grating (VIOLIN)

Deming Hu, Zhiyuan Xu, Junqiu Long, Peng Xiao, Lili Liang, Lipeng Sun, Hao Liang, Yang Ran * and Bai-Ou Guan

Guangdong Provincial Key Laboratory of Optical Fiber Sensing and Communications, Institute of Photonics Technology, Jinan University, Guangzhou 510632, China; hdming@stu2017.jnu.edu.cn (D.H.); 2016jnuxzy@stu2018.jnu.edu.cn (Z.X.); long@stu2018.jnu.edu.cn (J.L.); xiaopeng@stu2017.jnu.edu.cn (P.X.); lianglili@jnu.edu.cn (L.L.); lpsun@jnu.edu.cn (L.S.); tlianghao@jnu.edu.cn (H.L.); tguanbo@jnu.edu.cn (B.-O.G.)

* Correspondence: tranyang@jnu.edu.cn

Received: 28 May 2020; Accepted: 15 June 2020; Published: 17 June 2020

Abstract: Fiber optical refractometers have gained a substantial reputation in biological and chemical sensing domain regarding their label-free and remote-operation working mode. However, the practical breakthrough of the fiber optical bio/chemosensor is impeded by a lack of reconfigurability as well as the explicitness of the determination between bulk and surface refractive indices. In this letter, we further implement the highly flexible and reproducible long period grating called "VIOLIN" in chemical sensing area for the demonstration of moving those obstacles. In this configuration, the liquid is not only leveraged as the chemical carrier but also the periodic modulation of the optical fiber to facilitate the resonant signal. The thiol compound that is adsorbed by the fluidic substrate can be transduced to the pure alteration of the bulk refractive index of the liquid, which can be sensitively perceived by the resonant drift. Taking advantage of its freely dismantled feature, the VIOLIN sensor enables flexible reproduction and high throughput detection, yielding a new vision to the fiber optic biochemical sensing field.

Keywords: fiber optics; long period grating; fiber optical sensors; refractive index; chemical sensing; mercapto compound

---

1. Introduction

Optical fiber refractometer is a fast-developing candidate for biological and chemical sensing due to its promising feature of without being subjected to molecular dying-process and laboratory settings [1–4]. Various types of fiber optic refractive index sensors have been developed to pry about the ambient medium and can be categorized into gratings [5–7], interferometers [8–10], and resonators [11–14]. Among those sensors, long-period fiber grating (LPG) guides the core lightwave to outer cladding region via a longitudinal index modulation structure with a period of hundreds of microns to millimeters, allowing light to interplay further with the ambient medium. The LPG outperforms its counterparts lying in the combination of ease of fabrication, the flexibility of design, the abundance of serviceable signals, cost-effectiveness, and high sensitivity [15–17]. As a consequence, the LPG refractometer is soon considered as a competitive approach for label-free and in-situ assessing biological and chemical targets in which the molecular reaction events could be transduced simply by the grating resonance shift.

In the biosensing field, Chiavaioli et al. reported an IgG/Anti-IgG bioassay using the LPG with a copolymer functional layer [18]; Liu et al. utilized the graphene oxide functionalized LPG for the detection of the IgG [19] and hemoglobin [20], respectively; Xiao et al. facilitated a higher order diffraction LPG for the analysis of prostate specific antigen [21]; Piestrzyńska et al. proposed a tantalum oxide nano coated LPG for the test of avidin and *Escherichia coli* [22]; Yang et al. used a polyelectrolyte coated LPG to sense the *staphylococcus aureus* bacteria [23]; Janczuk-Richter et al. achieved the LPG-based the virus sensor [24] and Quero et al. realized the cancer biomarker detection upon applying the reflection mode of the LPG [25].

In chemosensing area, Yin et al. presented a 3D patterning of poly(acrylic acid) ionic hydrogel decorated LPG pH Sensor [26]; Wang et al. adopted LPG sensing approach to measure humidity [27]; Wang et al. carried out an LPG sensor with nano-assembled porphyrin layers to detect the concentration of ammonia gas [28]; Hsu et al. used a double notched LPG to sense CO_2 gas [29]; Baliyan et al. proposed a lipid sensor using the LPG [30]; Celebanska et al. and Tripathi et al. had conducted their trials on the aptasensor using LPG for monitoring cocaine [31] and toxin [32], respectively.

For those demonstrations, imposing a permanent index texture to the fiber is considered as a laboratory routine to make an LPG and the analytical solution commonly serves as the bulk milieu. In this strategy, reconfigurability and recyclability of the LPG sensor, however, remain challenging once the target molecules are immobilized specifically on the device, impeding the promotion of the LPG biochemical sensor with respect to the scenarios of high throughput screening as well as commercial Point-of-Care test (PoCT).

To overcome the limitations, a promising alternative was reported by our group through the use of a microfiber for the readout of information of a periodically patterned liquid, which can be described in principle as "Liquid renders the resonance to light. Light deciphers the secret of liquid" [33]. The vertical-fluid-array-induced optical microfiber long-period grating, analogous to a violin instrument, further harnesses the light-liquid interaction in a highly flexible and practical manner and facilitates the reproducibility through its nature of free combination of the two components, i.e., "fiber bow" and "fluidic pad".

In this letter, we further develop and demonstrate the VIOLIN sensor for the determination of chemicals. Mercapto compounds, including cysteine, homocysteine, and glutathione, play an essential role in the daily activity of lives. The alteration of the level of the mercapto compounds in bio-system may forecast some critical diseases, such as cancer or cardiovascular disease [34–38]. Here, a kind of mercapto compound, *p*-mercapto benzoic acid (*p*-MBA), was involved in the test. The channels of the VIOLIN, which are gilded by a thin gold layer, could adsorb the *p*-MBA molecules as the solution flows in the channels, resulting in a decrease of the bulk refractive index of the liquid and therefore a blueshift of the VIOLIN resonance. The amount of the resonant blueshift indicates the concentration of the *p*-MBA of the liquid quantitatively. Compared with the state-of-the-art of the LPG sensors in literature, this work fully takes advantage of the liquid-light interaction and the recombined feature of the VIOLIN structure. With high specificity and reconfigurability, this novel VIOLIN sensor would bear great potential and occupy a competitive niche in the label-free detection of chemical and biological molecules.

2. Materials and Methods

2.1. Materials

The optical fiber we used is the commercial single-mode telecom silica fiber (8/125 μm) with a numerical aperture (N.A.) of 0.14, which was obtained from Corning Inc (SMF-28, Corning, NY, USA). Rhodamine B (RhB) and p-Mercaptobenzoic acid (*pMBA*) were purchased from Macklin (Shanghai, China) and Aladdin (Shanghai, China), respectively. All the chemicals and reagents are at the highest purity grade available and used as received. The polymethyl methacrylate (PMMA) plate was gotten from Xintao Group (Shenzhen, China). The pure 100%-ethanol was acquired from HUSHI (Shanghai, China) and was used throughout the experiment.

2.2. Instrumentation

The CO_2 laser was purchased from SYNRAD with a home-made lasing-manipulation system (SYNRAD 48, Mukilteo, WA, USA). The gold-sputtering machine was the Ion Sputter Coater (SBC-12, KYKY Ltd., Beijing, China). A broadband LED source (GoLigtht Ltd., Shenzhen, China) was utilized to launch a continuous spectrum light ranging from 1250 to 1650 nm into the microfiber. An optical spectrum analyzer (OSA, AQ6370D, YOKOGAWA, Tokyo, Japan) was used for monitoring the output spectrum of the microfiber. A standard refractometer (PAL-RI, ATAGO, Tokyo, Japan) was used to calibrate the refractive index of the solution. A scanning electronic microscope (SEM Phenom pure+, Thermo Fisher Scientific, Eindhoven, The Netherlands) and a Raman Spectrometer (DXR3, Thermo Fisher Scientific, Waltham, MA, USA) were employed to characterize the substrates.

2.3. Principle of VIOLIN Device

The mode characteristics of an optical fiber could be fully elucidated through the finite-element-method (FEM) to solve a three-layer cylindrical waveguide with the step-change indices. For an LPG, the coupling between the fundamental mode and higher-order mode can be described according to the phase match condition as [39]:

$$|\beta 1 - \beta 2| = \frac{2\pi}{\Lambda} \tag{1}$$

where $\beta 1$ and $\beta 2$ are the propagating constants of the fundamental mode and one higher-order mode, respectively, and Λ refers to the grating period. The resonant wavelengths (λ_{res}) of an LPG can be expressed by transforming the Equation (1) to:

$$\lambda_{\text{res}} = \left(N_{eff\_o} - N_{eff\_v}\right) \times \Lambda = \Delta n \times \Lambda \tag{2}$$

where N_{eff_o} and N_{eff_v} represent the effective indices of the fundamental mode and the v-order mode, respectively. If the pattern of the grooves has a duty-cycle (Λ) of 1100 μm, a microfiber with a diameter of 35 μm is optimally selected for the design of an appropriate resonant wavelength within the telecommunication waveband. The inter-modal coupling occurs between the modes of LP01 and LP21 [33].

2.4. Configuration of the VIOLIN Sensor

Figure 1a shows the diagram of the VIOLIN structure, which is consist of two elements, "string pad"—The substrate and "bow"—The optical microfiber. First, the substrate of the VIOLIN takes the PMMA plate as the base that has dimensions of $(50 \times 30 \times 3)$ mm^3 (L × W × H). The 10.6 μm—Wavelength pulsed CO_2 laser is used to inscribe the V-groove-channels on the PMMA plate. The laser beam is focused to a spot of a diameter of 50 μm by a ZnSe lens. The scanning speed, output power, and repetition frequency of the CO_2 laser are set to 1000 mm/s, 25 W, and 8 kHz, respectively, which are precisely controlled by the home-made control system. Ten channels are neatly designed, as shown in Figure 1b. Each V-groove-channel is engraved through an 8 cycle/3 s -laser irradiation. The channel has a width of 360 μm and a depth of 760 μm, as seen in Figure 1c. The period of the channels is 1100 μm. After the inscription, the substrate is coated with a thin film of gold by the Ion Sputter Coater. Second, the optical microfiber is drawn from the commercial single-mode fiber using the flame-heated drawing technique [40,41]. The diameter of the taper waist of the microfiber is set to 35 μm and placed vertically to the V-groove-channel array, as shown in Figure 1d. The analytical liquids are pipetted into the channels for the activation of the VIOLIN. The broadband LED and OSA are connected with the microfiber by the two ends to monitor the transmission spectrum and thus facilitate the analytical readout. The experiment is carried out in the air-condition controlled lab and the environmental temperature is kept at 24 °C. We have tried to make all the experiments at almost the same room-temperature to eliminate the temperature cross-sensitivity [33].

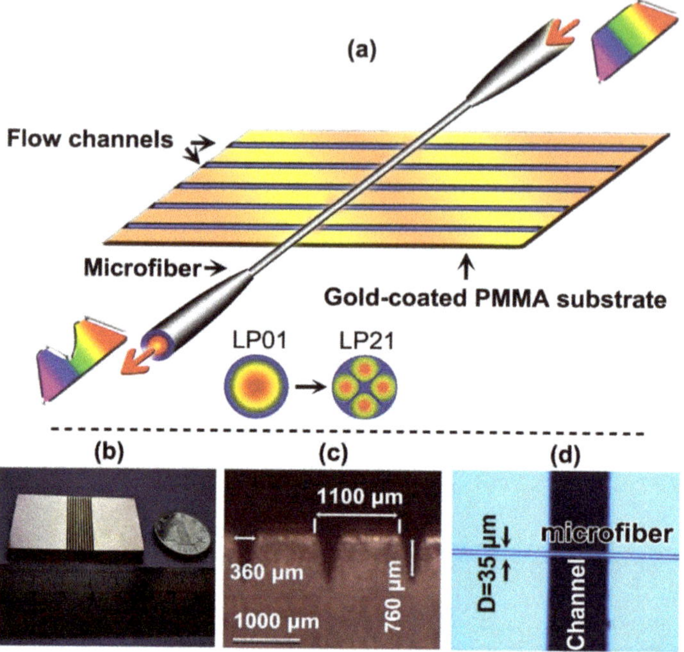

Figure 1. (**a**) Diagrams of the schematic of the VIOLIN sensor; Inset: the coupling between the two modes as indicated. (**b**) Real product of the substrate. (**c**) Scanning electron microscopic (SEM) image of the fluidic channel (horizontal view). (**d**) The optical microscopic image of the microfiber vertically placed on the substrate (perpendicular view).

3. Results

3.1. Refractive-Index Response of the VIOLIN

The label-free sensing often harnesses a suitable and reliable transducing parameter. In this configuration, since the liquid plays an indispensable role in the VIOLIN structure, solutes in the liquid would affect the resonant spectrum tremendously through the alteration of the refractive index. As a consequence, investigating the influence of changing different RI of liquids in the grooves is of great necessity. A series of liquids are prepared by mixing the pure ethanol and deionized water with different ratios. The concentrations of the ethanol in the mix were adjusted and represented by the medium RI. The RIs of the mix were calibrated by the refractometer, ranging from 1.336 to 1.361. Figure 2a displays the spectra of the resonances with respect to those liquids which are pipetted into the channels. As the RI of the fluid increases, the resonant wavelength moves to the longer wavelength. It is worth noting that variation of the spectral shape could be observed in the RI increasing process. The reason should be attributed to that the liquid was not flowing ideally in the groove channels with identical volume through manual operation of pipetting, probably leading to the insufficient interaction between the liquid and the microfiber at some channels. Nevertheless, the periodic structure could still guarantee a convincing resonant wavelength of the configuration for revealing the RI response. The response curve, with a linear correlation ($R^2 = 0.99$), outputs a sensitivity of ~2228 nm/RIU, which is similar to our previous report. The high RI sensitivity allows the VIOLIN to perceive a slight change in the concentration of the chemical molecules in the solution.

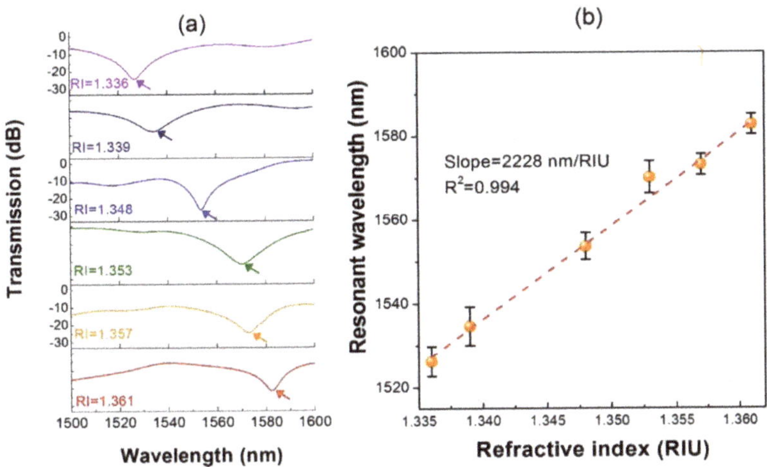

Figure 2. (**a**) Spectral change of the VIOLIN with respect to the RI increase of the liquid. (**b**) Response curve between the resonance wavelength and the liquid RI. Error bars indicate the standard deviations of three independent measurements and below were the same.

3.2. VIOLIN for Mercapto-Group Chemical Sensing

Since chemicals with mercapto-group tend to be adsorbed covalently to the metal surface [42–44], the VIOLIN can be utilized as the thiol group chemical sensors by virtue of the gold-gilded substrate. In this experiment, we employ the p-mercapto benzoic acid (*p*MBA) as the chemical which is dissolved into the pure ethanol with the concentration of 10^{-3} M. As the liquid is pipetted into the channels, the gold layer at the wall of the channel captures the *p*MBA molecules from the solution, leading to a lower density of the solution. Therefore, the resonance dip of the VIOLIN exhibits a blue-shifting curve along with time elapsing (150 s) due to the refractive index decrease during the adsorption process, as shown in the inset of Figure 3. The total wavelength shift is ~−12.5 nm, which can be obtained by the value subtraction between the plateaus from the end and the beginning. However, if the pure ethanol liquid without *p*MBA molecule is injected into the channels, the resonance maintains its spectral position as a reliable refractometer. By contrast, we test another VIOLIN without gilding the substrate to target the same *p*MBA solution. It can be seen in the Figure 3, gilding-free VIOLIN presents a significantly different response. The red-shift of the resonance probably indicates the weak binding between the *p*MBA and silica fiber via van der Waals force. Therefore, the gold layer in the channels of the VIOLIN is essential to facilitate mercapto-group chemical sensing.

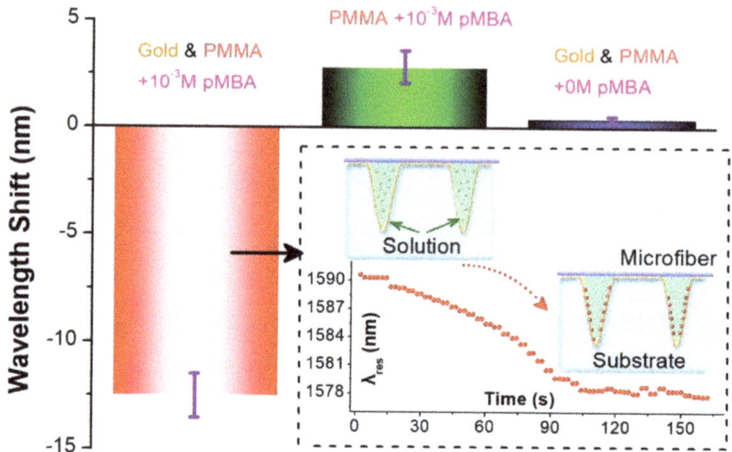

Figure 3. Response of the VIOLIN sensors for targeting *pMBA* solutions. Inset: the diagram and dynamic process of *pMBA* sensing using the gilded-VIOLIN.

To quantitatively investigate the sensing capability of the VIOLIN towards mercapto-group chemicals, we have prepared a series of concentrations of *p*MBA ethanol solutions ranging from 10^{-5} to 10^{-1} M. The liquids are pipetted into the channels in the sequential order of the concentrations from low to high. At each concentration, three solution samples are tested. The wavelength shift of the resonance ($\Delta\lambda$) is obtained by the subtraction between the wavelength values recorded at 120 s and the beginning of the process, respectively.

As illustrated in the Figure 4, the higher concentration of *p*MBA ethanol solution induces a larger response of the VIOLIN resonance as a result of more molecules are expelled from the solution and thus a more significant change of the refractive index. The curve follows a logistic fitting which is described as:

$$\Delta\lambda = -28.49 + \frac{27.85}{(1 + C/0.0016)^{0.763}}, \left(R^2 = 0.976\right) \quad (3)$$

where C denotes the concentration of the *p*MBA. Moreover, a log-linear responding region was found with the range from 10^{-4} to 10^{-1} M. The relationship, shown in the inset of Figure 4, could be represented as

$$\Delta\lambda = -8.22 \times log(C) - 36.233, \left(R^2 = 0.993\right) \quad (4)$$

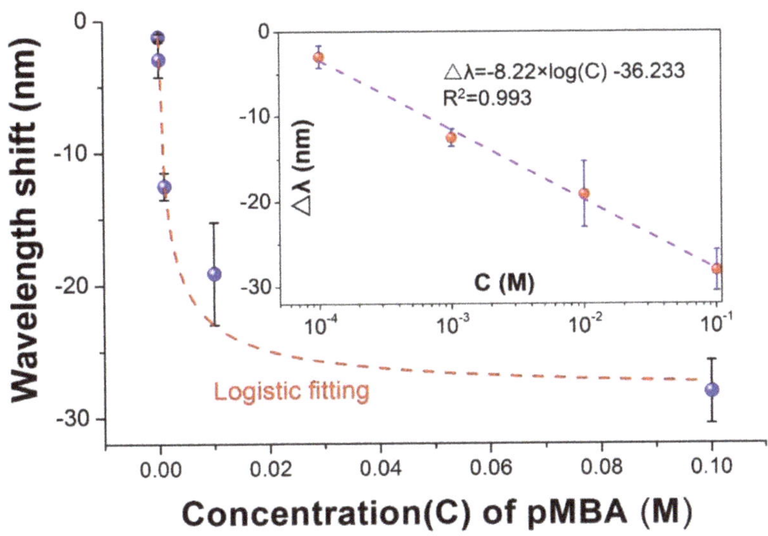

Figure 4. Response curve of sensing *p*MBA solutions with different concentrations using VIOLIN chemical sensor. Inset: the log-linear correlation at part region of the concentrations.

To verify the adsorption of the *p*MBA on the surface of the groove, we use a Raman Spectrometer to analyze the Raman scattering spectrum by focusing the groove channel after several rounds of washing and air-drying to eliminate the non-covalently bonded residuals. From Figure 5, we can see that the *p*MBA ethanol solution of higher concentration would enhance the Raman peaks of 1078 cm^{-1} and 1580 cm^{-1}, which are the "fingerprint" spectra of *p*MBA due to the ν12 and ν8a planar vibrations of the benzene ring, respectively [45]. The result confirms that the gold layer in the groove captures the *p*MBA in the solution effectively and therefore reduces the refractive index of the liquid on a relatively large scale.

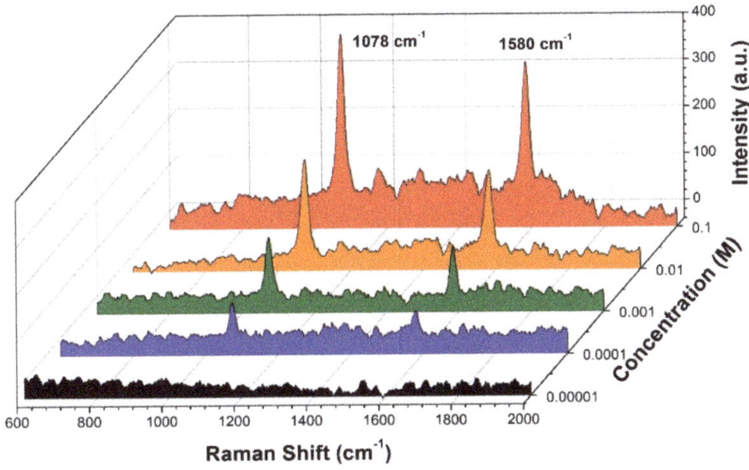

Figure 5. Raman intensities of the dried groove channel after the flowing of the *pMBA* ethanol solutions.

3.3. Contrast Response to the Non-Mercapto-Group Chemicals

In order to prove the specificity of the VIOLIN, we employ the ethanol solution containing Rhodamine B, a kind of non-thiol chemical, in the experiment [46]. After pipetting the different concentrations of RhB ethanol solutions in the channel array, it can be seen in Figure 6 that the wavelength-shifting of the VIOLIN resonance exhibits irregular performance in spite of the concentration changing of the analytes. As well, the amounts of the wavelength shifts are less than 2 nm at different concentrations and much smaller in contrast with the *p*MBA test even at a moderate concentration of the solution (12 nm @10^{-3} M). The erratic and tiny response indicates the unaltered concentration of the solution in the channels due to that the molecular binding neither occurs at substates surface nor the fiber surface. The result manifests that the VIOLIN holds the specificity for targeting mercapto-group chemicals.

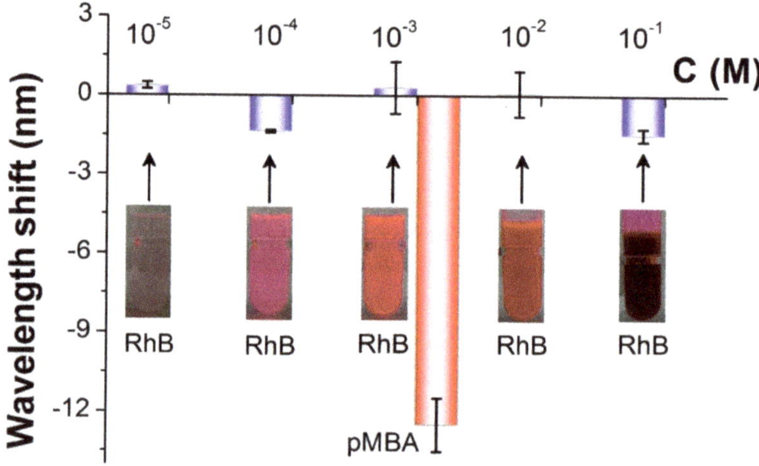

Figure 6. The response of the VIOLIN to different RhB ethanol solutions in comparison with the response to the moderate concentration of *p*MBA. "C" referred to "concentration".

3.4. Reproducibility of the VIOLIN Chemical Sensor

In the chemical or biological sensing area, recycling of the sensor, commonly applying "probe-target" molecularly specific binding mechanism, is often complicated due to the additional unbinding process. The VIOLIN configuration offers an excellent solution thanks to its unique superiority of flexible reconstruction of the components. The microfiber and substrate could be dissembled and assembled freely as though we are playing the bow and string pad from different real violin instruments. The convenience of the reproduction of the VIOLIN could be demonstrated via the changing of the substrates. The CO_2 laser engraving technique provides high reproducibility in manufacturing and throughput to the substrates. Three of the substrates are randomly selected from a batch with the specified engraving parameters as mentioned above. The same microfiber is placed on those substrates in turns. The same concentration of 0.1 M *p*MBA ethanol solution is used throughout the tests. The results, shown in Figure 7, describe that the three VIOLINs present similar responses of blue-shift, −31 nm on average with a standard deviation of 2.5 nm, to the *p*MBA ethanol solution. Therefore, the detachable structure enables the proposed VIOLIN to revive by changing the substrates after an irreversible binding detection (*p*MBA molecules—Gold).

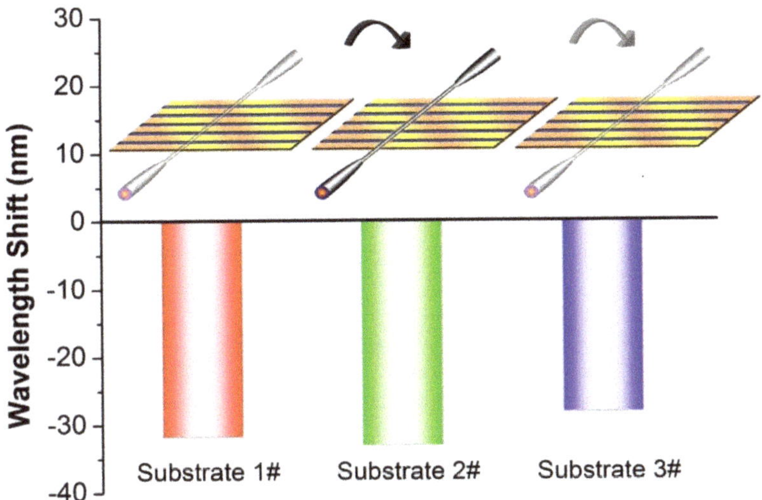

Figure 7. Response of the VIOLINs, using the substrates in a batch, with respect to 0.1 M *pMBA* solution. Inset: the scheme of substate-changing in the tests.

4. Conclusions

In summary, a highly reconfigurable and scalable long period fiber grating (VIOLIN) is demonstrated to be a fascinating analytical approach of the mercapto group chemicals. The liquids streaming in the periodic groove channels are devoted to the LPG as the modulating structure acting on the microfiber. The VIOLIN presents a high RI sensitivity over 2000 nm/RIU, allowing for perceiving the density change of the liquid caused by the precipitation of the solute. By gilding a gold layer, the VIOLIN is endowed with the capability of sensing mercapto group chemicals with high specificity compared with non-mercapto group chemicals. The response of the resonance purely depends on the alteration of the bulk refractive index as a result of the adsorption of the *pMBA* on the gold layer, which is confirmed by the Raman analysis in the groove channel. It is a paradigm shift to the traditional surface refractive index transducing regime, enabling an explicit and straightforward assay approach. Besides, the configuration offers high flexibility and reproducibility to the formation of LPG. The free detachability enables the VIOLIN to be reproduced by merely changing the substrate after an irreversible adsorption process without the laborious operation of applying a new fiber sensor. Future investigation should be focused on the surfactant modification in the groove to address the limitation of the aqueous fluidity caused by hydrophilia for expanding the application scenarios, especially for the case of the determination of biomolecules. Furthermore, programmable manipulation of the fluidic pipetting can be involved in tailoring the VIOLIN spectrum from several aspects, such as linewidth, chirping as well as phase-shifting, to provide more functionalities. Besides, several approaches of temperature compensating or simultaneous monitoring could be involved to overcome the temperature cross-sensitivity. It is predicted that the multiplexing of the VIOLIN could contribute to the temperature-RI simultaneous sensing by flowing different types of liquid with distinct thermo-optic coefficients. Overall, thanks to the high RI sensitivity, facility, reproducibility, and scalability, the sensor proposed in our scheme offers a new route to the field of chemical and biological sensing through catering to the need of high throughput test and commercialization.

Author Contributions: Y.R., and B.-O.G. proposed the idea; Y.R. and D.H. investigated the state-of-the-art of the topic and designed the experiment; D.H., Z.X., J.L., P.X., and L.L. collaborated in accomplishing the experiment; L.S. and H.L. provided the necessary suggestions in the experimental design and manuscript preparation; D.H. prepared the original draft; Y.R., and B.-O.G. wrote, reviewed and edited the manuscript; Y.R., and B.-O.G. provided the project administration. All authors have read and agreed to the published version of the manuscript.

Funding: This research was funded by the National Natural Science Foundation of China (61775082, U1701268, 61675091, 61705083, 61805106), the Local Innovative and Research Teams Project of Guangdong Pearl River Talents Program (2019BT02 × 105), Guangdong Natural Science Foundation (2018A030313677), and the Fundamental Research Funds for the Central Universities.

Conflicts of Interest: The authors declare no conflict of interest.

References

1. Chiavaioli, F.; Gouveia, A.J.C.; Jorge, A.S.P.; Baldini, F. Towards a Uniform Metrological Assessment of Grating-Based Optical Fiber Sensors: From Refractometers to Biosensors. *Biosensors* **2017**, *7*, 23. [CrossRef] [PubMed]
2. Guan, B.-O.; Huang, Y. Interface Sensitized Optical Microfiber Biosensors. *J. Lightwave Technol.* **2019**, *37*, 2616–2622. [CrossRef]
3. Socorro-Leránoz, A.B.; Santano, D.; Del Villar, I.; Matias, I.R. Trends in the design of wavelength-based optical fibre biosensors (2008–2018). *Biosens. Bioelectron. X* **2019**, *1*, 100015. [CrossRef]
4. Wang, X.-D.; Wolfbeis, O.S. Fiber-Optic Chemical Sensors and Biosensors (2015–2019). *Anal. Chem.* **2020**, *92*, 397–430. [CrossRef] [PubMed]
5. Guo, T.; Liu, F.; Liu, Y.; Chen, N.-K.; Guan, B.-O.; Albert, J. In-situ detection of density alteration in non-physiological cells with polarimetric tilted fiber grating sensors. *Biosens. Bioelectron.* **2014**, *55*, 452–458. [CrossRef]
6. Biswas, P.; Chiavaioli, F.; Jana, S.; Basumallick, N.; Trono, C.; Giannetti, A.; Tombelli, S.; Mallick, A.; Baldini, F.; Bandyopadhyay, S. Design, fabrication and characterisation of silica-titania thin film coated over coupled long period fibre gratings: Towards bio-sensing applications. *Sens. Actuators B Chem.* **2017**, *253*, 418–427. [CrossRef]
7. Liu, T.; Liang, L.-L.; Xiao, P.; Sun, L.-P.; Huang, Y.-Y.; Ran, Y.; Jin, L.; Guan, B.-O. A label-free cardiac biomarker immunosensor based on phase-shifted microfiber Bragg grating. *Biosens. Bioelectron.* **2018**, *100*, 155–160. [CrossRef]
8. Li, Y.; Tong, L. Mach-Zehnder interferometers assembled with optical microfibers or nanofibers. *Opt. Lett.* **2008**, *33*, 303–305. [CrossRef]
9. Li, J.; Sun, L.-P.; Gao, S.; Quan, Z.; Chang, Y.-L.; Ran, Y.; Jin, L.; Guan, B.-O. Ultrasensitive refractive-index sensors based on rectangular silica microfibers. *Opt. Lett.* **2011**, *36*, 3593–3595. [CrossRef]
10. Sun, L.-P.; Li, J.; Tan, Y.; Gao, S.; Jin, L.; Guan, B.-O. Bending effect on modal interference in a fiber taper and sensitivity enhancement for refractive index measurement. *Opt. Express* **2013**, *21*, 26714–26720. [CrossRef]
11. Sumetsky, M.; Dulashko, Y.; Fini, J.M.; Hale, A. Optical microfiber loop resonator. *Appl. Phys. Lett.* **2005**, *86*, 161108. [CrossRef]
12. Jiang, X.; Chen, Y.; Vienne, G.; Tong, L. All-fiber add-drop filters based on microfiber knot resonators. *Opt. Lett.* **2007**, *32*, 1710–1712. [CrossRef] [PubMed]
13. Xu, F.; Brambilla, G. Demonstration of a refractometric sensor based on optical microfiber coil resonator. *Appl. Phys. Lett.* **2008**, *92*, 101126. [CrossRef]
14. Zubiate, P.; Urrutia, A.; Zamarreño, C.R.; Egea-Urra, J.; Fernández-Irigoyen, J.; Giannetti, A.; Baldini, F.; Díaz, S.; Matias, I.R.; Arregui, F.J.; et al. Fiber-based early diagnosis of venous thromboembolic disease by label-free D-dimer detection. *Biosens. Bioelectron. X* **2019**, *2*, 100026. [CrossRef]
15. Pan, Z.; Huang, Y.; Xiao, H. Multi-Parameter Sensing Device to Detect Liquid Layers Using Long-Period Fiber Gratings. *Sensors* **2018**, *18*, 3094. [CrossRef]
16. Janik, M.; Koba, M.; Król, K.; Mikulic, P.; Bock, J.W.; Śmietana, M. Combined Long-Period Fiber Grating and Microcavity In-Line Mach–Zehnder Interferometer for Refractive Index Measurements with Limited Cross-Sensitivity. *Sensors* **2020**, *20*, 2431. [CrossRef]
17. Torres-Gómez, I.; Ceballos-Herrera, E.D.; Salas-Alcantara, M.K. Mechanically-Induced Long-Period Fiber Gratings Using Laminated Plates. *Sensors* **2020**, *20*, 2582. [CrossRef]

18. Chiavaioli, F.; Trono, C.; Giannetti, A.; Brenci, M.; Baldini, F. Characterisation of a label-free biosensor based on long period grating. *J. Biophotonics* **2014**, *7*, 312–322. [CrossRef]
19. Liu, C.; Cai, Q.; Xu, B.; Zhu, W.; Zhang, L.; Zhao, J.; Chen, X. Graphene oxide functionalized long period grating for ultrasensitive label-free immunosensing. *Biosens. Bioelectron.* **2017**, *94*, 200–206. [CrossRef]
20. Liu, C.; Xu, B.J.; Zhou, L.; Sun, Z.; Mao, H.J.; Zhao, J.L.; Zhang, L.; Chen, X. Graphene oxide functionalized long period fiber grating for highly sensitive hemoglobin detection. *Sens. Actuators B Chem.* **2018**, *261*, 91–96. [CrossRef]
21. Xiao, P.; Sun, Z.; Huang, Y.; Lin, W.; Ge, Y.; Xiao, R.; Li, K.; Li, Z.; Lu, H.; Yang, M.; et al. Development of an optical microfiber immunosensor for prostate specific antigen analysis using a high-order-diffraction long period grating. *Opt. Express* **2020**, *28*, 15793. [CrossRef]
22. Piestrzyńska, M.; Dominik, M.; Kosiel, K.; Janczuk-Richter, M.; Szot-Karpińska, K.; Brzozowska, E.; Shao, L.; Niedziółka-Jonsson, J.; Bock, W.J.; Śmietana, M. Ultrasensitive tantalum oxide nano-coated long-period gratings for detection of various biological targets. *Biosens. Bioelectron.* **2019**, *133*, 8–15. [CrossRef] [PubMed]
23. Yang, F.; Chang, T.-L.; Liu, T.; Wu, D.; Du, H.; Liang, J.; Tian, F. Label-free detection of Staphylococcus aureus bacteria using long-period fiber gratings with functional polyelectrolyte coatings. *Biosens. Bioelectron.* **2019**, *133*, 147–153. [CrossRef] [PubMed]
24. Janczuk-Richter, M.; Dominik, M.; Roźniecka, E.; Koba, M.; Mikulic, P.; Bock, W.J.; Łoś, M.; Śmietana, M.; Niedziółka-Jönsson, J. Long-period fiber grating sensor for detection of viruses. *Sens. Actuators B Chem.* **2017**, *250*, 32–38. [CrossRef]
25. Quero, G.; Consales, M.; Severino, R.; Vaiano, P.; Boniello, A.; Sandomenico, A.; Ruvo, M.; Borriello, A.; Diodato, L.; Zuppolini, S.; et al. Long period fiber grating nano-optrode for cancer biomarker detection. *Biosens. Bioelectron.* **2016**, *80*, 590–600. [CrossRef] [PubMed]
26. Yin, M.-J.; Yao, M.; Gao, S.; Zhang, A.P.; Tam, H.-Y.; Wai, P.-K.A. Rapid 3D Patterning of Poly(acrylic acid) Ionic Hydrogel for Miniature pH Sensors. *Adv. Mater.* **2016**, *28*, 1394–1399. [CrossRef]
27. Wang, Y.; Liu, Y.; Zou, F.; Jiang, C.; Mou, C.; Wang, T. Humidity Sensor Based on a Long-Period Fiber Grating Coated with Polymer Composite Film. *Sensors* **2019**, *19*, 2263. [CrossRef]
28. Wang, T.; Yasukochi, W.; Korposh, S.; James, S.W.; Tatam, R.P.; Lee, S.-W. A long period grating optical fiber sensor with nano-assembled porphyrin layers for detecting ammonia gas. *Sens. Actuators B Chem.* **2016**, *228*, 573–580. [CrossRef]
29. Hsu, H.-C.; Hsieh, T.-S.; Huang, T.-H.; Tsai, L.; Chiang, C.-C. Double Notched Long-Period Fiber Grating Characterization for CO2 Gas Sensing Applications. *Sensors* **2018**, *18*, 3206. [CrossRef]
30. Baliyan, A.; Sital, S.; Tiwari, U.; Gupta, R.; Sharma, E.K. Long period fiber grating based sensor for the detection of triacylglycerides. *Biosens. Bioelectron.* **2016**, *79*, 693–700. [CrossRef]
31. Celebanska, A.; Chiniforooshan, Y.; Janik, M.; Mikulic, P.; Sellamuthu, B.; Walsh, R.; Perreault, J.; Bock, W.J. Label-free cocaine aptasensor based on a long-period fiber grating. *Opt. Lett.* **2019**, *44*, 2482–2485. [CrossRef]
32. Tripathi, S.M.; Dandapat, K.; Bock, W.J.; Mikulic, P.; Perreault, J.; Sellamuthu, B. Gold coated dual-resonance long-period fiber gratings (DR-LPFG) based aptasensor for cyanobacterial toxin detection. *Sens. Bio-Sens. Res.* **2019**, *25*, 100289. [CrossRef]
33. Ran, Y.; Hu, D.; Xu, Z.; Long, J.; Guan, B.-O. Vertical-fluid-array induced optical microfiber long period grating (VIOLIN) refractometer. *J. Lightwave Technol.* **2020**. [CrossRef]
34. Wu, J.; Sheng, R.; Liu, W.; Wang, P.; Ma, J.; Zhang, H.; Zhuang, X. Reversible Fluorescent Probe for Highly Selective and Sensitive Detection of Mercapto Biomolecules. *Inorg. Chem.* **2011**, *50*, 6543–6551. [CrossRef]
35. Miller, J.W.; Beresford, S.A.A.; Neuhouser, M.L.; Cheng, T.-Y.D.; Song, X.; Brown, E.C.; Zheng, Y.; Rodriguez, B.; Green, R.; Ulrich, C.M. Homocysteine, cysteine, and risk of incident colorectal cancer in the Women's Health Initiative observational cohort. *Am. J. Clin. Nutr.* **2013**, *97*, 827–834. [CrossRef] [PubMed]
36. Lan, M.; Zhang, J.; Chui, Y.-S.; Wang, H.; Yang, Q.; Zhu, X.; Wei, H.; Liu, W.; Ge, J.; Wang, P.; et al. A recyclable carbon nanoparticle-based fluorescent probe for highly selective and sensitive detection of mercapto biomolecules. *J. Mater. Chem. B* **2015**, *3*, 127–134. [CrossRef]
37. Liu, P.; Qi, C.-B.; Zhu, Q.-F.; Yuan, B.-F.; Feng, Y.-Q. Determination of thiol metabolites in human urine by stable isotope labeling in combination with pseudo-targeted mass spectrometry analysis. *Sci. Rep.* **2016**, *6*, 21433. [CrossRef]
38. Stachniuk, J.; Kubalczyk, P.; Furmaniak, P.; Głowacki, R. A versatile method for analysis of saliva, plasma and urine for total thiols using HPLC with UV detection. *Talanta* **2016**, *155*, 70–77. [CrossRef]

39. Xuewen, S.; Lin, Z.; Bennion, I. Sensitivity characteristics of long-period fiber gratings. *J. Lightwave Technol.* **2002**, *20*, 255–266. [CrossRef]
40. Ran, Y.; Tan, Y.-N.; Sun, L.-P.; Gao, S.; Li, J.; Jin, L.; Guan, B.-O. 193 nm excimer laser inscribed Bragg gratings in microfibers for refractive index sensing. *Opt. Express* **2011**, *19*, 18577–18583. [CrossRef]
41. Ran, Y.; Jin, L.; Sun, L.P.; Li, J.; Guan, B.O. Temperature-Compensated Refractive-Index Sensing Using a Single Bragg Grating in an Abrupt Fiber Taper. *IEEE Photonics J.* **2013**, *5*, 7100208.
42. Lou, Z.; Han, H.; Zhou, M.; Wan, J.; Sun, Q.; Zhou, X.; Gu, N. Fabrication of Magnetic Conjugation Clusters via Intermolecular Assembling for Ultrasensitive Surface Plasmon Resonance (SPR) Detection in a Wide Range of Concentrations. *Anal. Chem.* **2017**, *89*, 13472–13479. [CrossRef] [PubMed]
43. Lou, Z.; Han, H.; Mao, D.; Jiang, Y.; Song, J. Qualitative and Quantitative Detection of PrPSc Based on the Controlled Release Property of Magnetic Microspheres Using Surface Plasmon Resonance (SPR). *Nanomaterials* **2018**, *8*, 107. [CrossRef] [PubMed]
44. Strobbia, P.; Ran, Y.; Crawford, B.M.; Cupil-Garcia, V.; Zentella, R.; Wang, H.-N.; Sun, T.-P.; Vo-Dinh, T. Inverse Molecular Sentinel-Integrated Fiberoptic Sensor for Direct and in Situ Detection of miRNA Targets. *Anal. Chem.* **2019**, *91*, 6345–6352. [CrossRef]
45. Smith, G.; Girardon, J.-S.; Paul, J.-F.; Berrier, E. Dynamics of a plasmon-activated p-mercaptobenzoic acid layer deposited over Au nanoparticles using time-resolved SERS. *Phys. Chem. Chem. Phys.* **2016**, *18*, 19567–19573. [CrossRef]
46. Ran, Y.; Strobbia, P.; Cupil-Garcia, V.; Vo-Dinh, T. Fiber-optrode SERS probes using plasmonic silver-coated gold nanostars. *Sens. Actuators B Chem.* **2019**, *287*, 95–101. [CrossRef]

MDPI

Article

Development of Real-Time Time Gated Digital (TGD) OFDR Method and Its Performance Verification

Kinzo Kishida [1,*], Artur Guzik [1], Ken'ichi Nishiguchi [1], Che-Hsien Li [1], Daiji Azuma [1], Qingwen Liu [2] and Zuyuan He [2]

1 Neubrex Co., Ltd. 1-1-24 Sakaemachi-dori, Kobe 650-0024, Japan; guzik@neubrex.jp (A.G.); nishiguchi@neubrex.jp (K.N.); li-z@neubrex.jp (C.-H.L.); azuma@neubrex.jp (D.A.)

2 State Key Laboratory of Advanced Optical Communicate Systems and Networks, Shanghai Institute for Advanced Communication and Data Science, Shanghai Jiao Tong University, 800 Dongchuan Rd., Shanghai 200240, China; liuqingwen@sjtu.edu.cn (Q.L.); zuyuanhe@sjtu.edu.cn (Z.H.)

* Correspondence: kishida@neubrex.jp

Abstract: Distributed acoustic sensing (DAS) in optical fibers detect dynamic strains or sound waves by measuring the phase or amplitude changes of the scattered light. This contrasts with other distributed (and more conventional) methods, such as distributed temperature (DTS) or strain (DSS), which measure quasi-static physical quantities, such as intensity spectrum of the scattered light. DAS is attracting considerable attention as it complements the conventional distributed measurements. To implement DAS in commercial applications, it is necessary to ensure a sufficiently high signal-noise ratio (SNR) for scattered light detection, suppress its deterioration along the sensing fiber, achieve lower noise floor for weak signals and, moreover, perform high-speed processing within milliseconds (or sometimes even less). In this paper, we present a new, real-time DAS, realized by using the time gated digital-optical frequency domain reflectometry (TGD-OFDR) method, in which the chirp pulse is divided into overlapping bands and assembled after digital decoding. The developed prototype NBX-S4000 generates a chirp signal with a pulse duration of 2 μs and uses a frequency sweep of 100 MHz at a repeating frequency of up to 5 kHz. It allows one to detect sound waves at an 80 km fiber distance range with spatial resolution better than a theoretically calculated value of 2.8 m in real time. The developed prototype was tested in the field in various applications, from earthquake detection and submarine cable sensing to oil and gas industry applications. All obtained results confirmed effectiveness of the method and performance, surpassing, in conventional SM fiber, other commercially available interrogators.

Keywords: OFDR type DAS; phase fading solution; high SNR; real-time events detection

Citation: Kishida, K.; Guzik, A.; Nishiguchi, K.; Li, C.-H.; Azuma, D.; Liu, Q.; He, Z. Development of Real-Time Time Gated Digital (TGD) OFDR Method and Its Performance Verification. *Sensors* **2021**, *21*, 4865. https://doi.org/10.3390/s21144865

Academic Editors: Gabriela Kuncová and Agostino Iadicicco

Received: 4 May 2021
Accepted: 8 July 2021
Published: 16 July 2021

1. Introduction

More than a decade ago, distributed acoustic sensing (DAS) became a popular measurement method, especially in the oil and gas industry [1]. In principle, the phase change between an incident light pulse and the returned Rayleigh scattered lights [1] is measured and determined, and using those data to detect and locate acoustic events. Typical applications of DAS include also seismic profiling, hydraulic fracturing monitoring, and intrusion detection. Those applications and corresponding technical requirements, which need to be met, are listed in Table 1 [2]. The first of the applications listed there, seismic wave measurements, has almost no real time requirements regarding processing time. In the other two applications, the size of the datum is in the order of terabytes per day, but it contains only a small amount of the necessary information. The necessary information is required to be output in real time.

Almost all DAS implementations, so far, are based on the optical time domain reflectometry (OTDR) technique [1]. Moreover, some of the authors of this paper had developed an OTDR-type interrogator, which used phase polarization diversity [3] to improve signal quality and reduce noise. The OTDR method has inherent problems, such as low sensitivity of acoustic measurements, Rayleigh scattering phase fading phenomena, and noise components of the system in the frequency domain contaminating the results. Despite that, some excellent results have been published by the industry; in each case, special methods were developed to deal with those OTDR related problems.

Table 1. Typical application fields and technical requirements for DAS.

Application	Technical Requirements
Earthquake measurements, underwater wave survey.	Recording of waves for a few seconds, time and location accuracy are important.
Hydraulic fracturing (shale gas).	It is necessary to display the processed results on the screen every few seconds during several days of continuous operation. Low frequency is important.
Events detection and classification.	Expectations for an intelligent society through the introduction of AI technology.

The frequency counterpart of OTDR, the optical frequency domain reflectometry (OFDR) type methods were also formulated. For those, time-gated digital (TGD)-OFDR, which applies a time gate to the chirp signal, has been studied [4,5]. The TGD method, when used for digital analysis, demonstrated excellent performance in the laboratory [4]. This paper describes the effort undertaken to develop a commercial level interrogator, as well as verification of its performance in real, in-the-field applications. It also contains and discusses obtained results and compares them with other works.

The TGD-DAS technology uses long optical pulses to solve the signal SNR problem. Chirp's pulse compression method can achieve a high spatial resolution of about 100 times. We also solved the phase-fading problem by dividing the frequency band of the chirp. Thus, TGD-DAS is expected to be the most promising DAS technology. During the presentation of the sensing method, SEAFOM [6] naming and technical terms are used.

2. System Configuration

The first (proof-of-concept measurements reported in [7]) were made on a laboratory system with a single data channel, capable of acquiring signals for the maximum duration of only a few seconds.

During the development described in this paper, we designed the system to meet three targets: being deployable in the field (commercial operational level), being capable of real-time data processing and, finally, and most importantly, to obtain signal in standard single-mode (SM) fibers of quality matching, or surpassing that with modified (referred to as engineered) fibers. The result of the system development is presented in Figure 1, which shows the configuration of the DAS prototype NBX-S4000 using TGD-OFDR.

As the light source, the fiber laser with a narrow linewidth (~100 Hz) is used. The chirp waveform is generated using the Acousto-Optic Modulator (AOM) of 200 MHz with the resulting chirp frequency range 145 MHz to 245 MHz. The signal is then acquired on a polarization diversity component and analyzed separately for both polarization components, P and S. Please note that the polarization components are denoted here by convention used in optics, and are not related to the P- and S-waves as used in seismology and other applications.

An example of a chirp signal (specifically its P-polarization component), as recorded in the NBX-S4000 instrument, is presented in Figure 2. We should note that this signal shown in this figure is obtained by connecting to the output fiber in the heterodyne receiver in the system. As a result, due to local reference light, the high frequency of amplitude is observed.

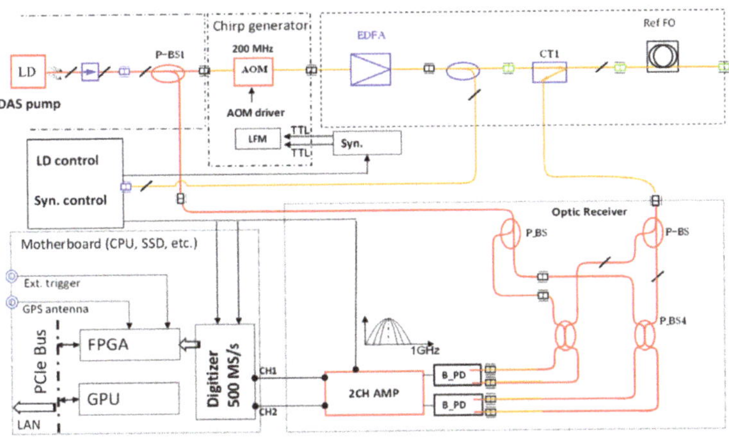

Figure 1. Configuration of prototype DAS NBX-S4000.

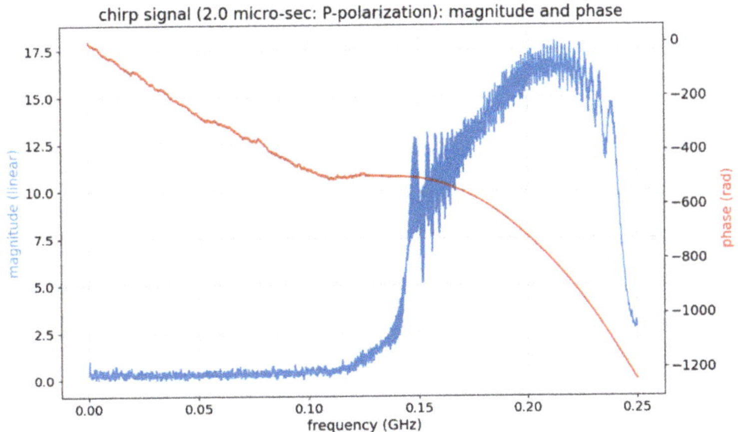

Figure 2. Chirp signal (P-polarization component) in NBX-S4000.

Once the signal is acquired and digitized, the signal processing is performed. The algorithm for signal processing is the same as already presented in [4,5], so it will not be repeated here. The resulting reflection coefficient trace is complex, and the magnitude part represents the reflection intensity trace of the measured fiber.

Three signal bands, using three decoding filters for each polarization component, are extracted from the acquired signal. This is schematically presented in Figure 3a. In this figure, the band frequencies are offset for the sake of readability, to indicate that they overlap. The normalized magnitudes of the bands in dB are shown in Figure 3b, where each band is indicated by a corresponding (matching) color.

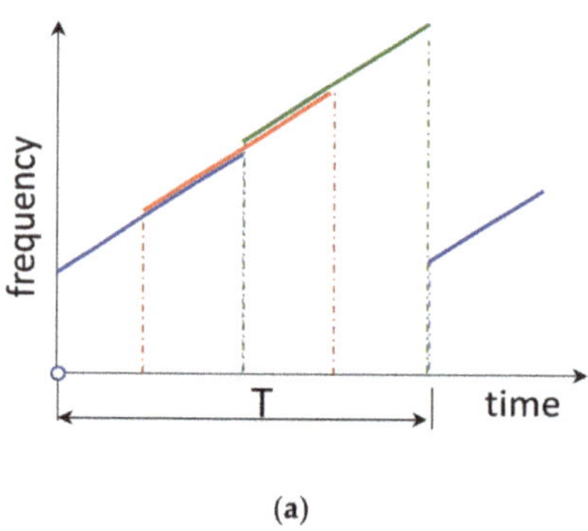

(a)

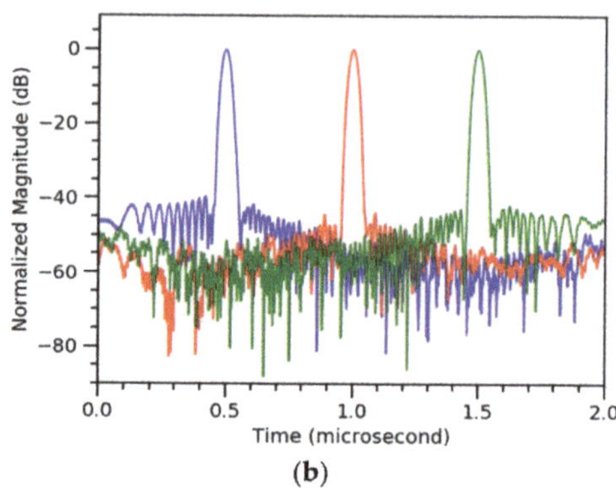

(b)

Figure 3. Signal decoding scheme. (**a**) Chirped pulse and sub-division into bands (offset for visibility, frequency change is linear over entire range); (**b**) corresponding frequency bands used during signal processing.

The decoding step, described above, is performed on an integrated, field-programmable gate array (FPGA). When used with a specially developed and optimized processing algorithm, FPGA makes it possible to obtain results in real-time, despite processing six data components (three bands for each polarization).

In many DAS systems, the actual complexity of the signal processing depends mainly on the part handling the occurrences of the phenomena referred to as phase fading and signal-to-noise ratio (SNR), in general. The achieved signal quality on the developed interrogator unit is examined in detail in the next section.

3. Signal-to-Noise Ratio (SNR)

The phase shift of the backscattered light is linearly proportional to the vibration/acoustic amplitude. However, along the fiber there exist (many) randomly distributed locations, where the intensity of backscattering is extremely low. This effect is mainly caused by interference fading or by polarization fading [3,4].

As the SNR is low in general, the extracted phase at these fading points contain large noise, making estimation and actual phase change calculations extremely difficult. An efficient method to reduce the fading noise is to vary the frequency of the probe light and average independent Rayleigh backscattering traces. In TGD-DAS, this is achieved by dividing the (chirp) signal into (overlapping) bands. Each band is then individually processed.

Figure 4 shows an example of the signal-to-noise ratio (SNR) distribution along the measured fiber. SNR is the ratio of the signal power to the noise power that corrupts the signal. In Figure 4, SNR is plotted separately, not only for each of the polarization components (Figure 4a,b), but also for individual processing frequency bands, denoted as P0, P1, P2, for P-polarization and S0, S1, and S2 for the S-polarization component. It can be observed that there are places along the fiber, which exhibit extremely low (negative) SNR values. If the datum from the individual band was only available, it would still be extremely challenging to obtain phase values at those phase fading locations. However, at any given location, if there is low SNR for one of the bands, then for the other bands—as they use other portions of the signal spectrum—SNR is high. Fortunately, the bands can (and are) combined, which means the fading phase location is virtually removed from the decoded signal. The SNR for combined bands for each of the polarization components are shown in Figure 5.

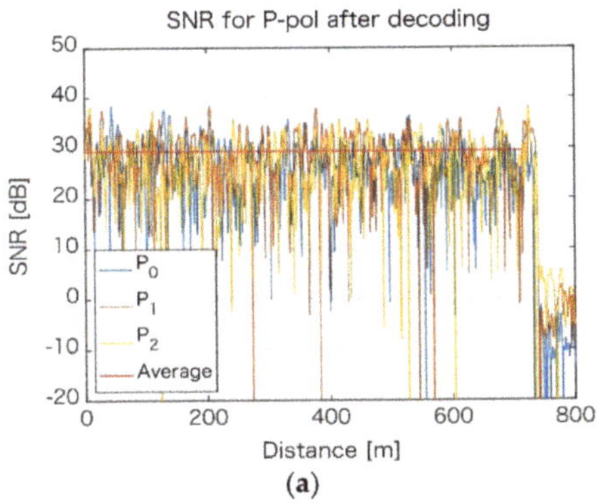

(a)

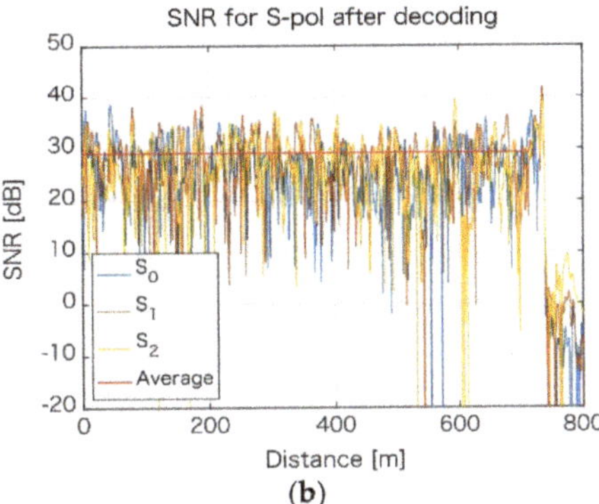

(b)

Figure 4. Signal-to-noise distribution for individual frequency bands. (**a**) P-polarization component; (**b**) S-polarization component.

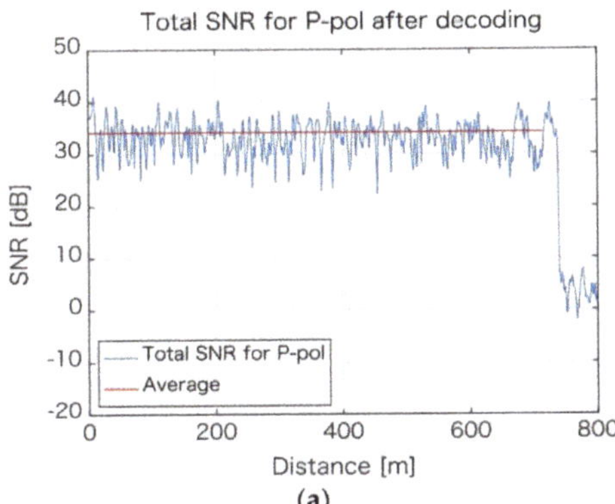

(a)

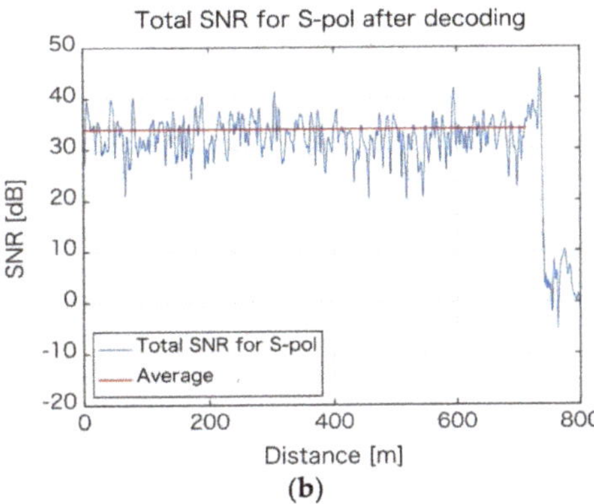

(b)

Figure 5. Signal-to-noise distribution along the fiber for P-polarization component (**a**) and S-polarization (**b**).

To make the situation even better, the data for acquired polarization components are also combined to yield results, as shown in Figure 6. Along the entire length of the fiber, there is no location where phase fading could be observed, and the lower SNR is approximately +30 dB, with an average of +38 dB.

When combining the signal from individual components, in any measurement system, one should consider and determine the influence of the system noise contributing to (and influencing) the minimum measurable level, referred to as the inherent measurement system noise floor. This important aspect is discussed next.

Noise Floor

For determining the noise floor of NBX-S4000, a typical procedure, in distributed optical fiber sensing and acoustic type of measurements, was used. First, the signal was acquired in a "non-acoustic" (also called "silent") portion of the fiber. Next, the full spectral analysis of the signal performed, resulting in obtaining power spectrum density (PSD) of the signal, which in communication is also referred to as noise spectral density, or noise power density. The unit of PSD is power of noise over the frequency.

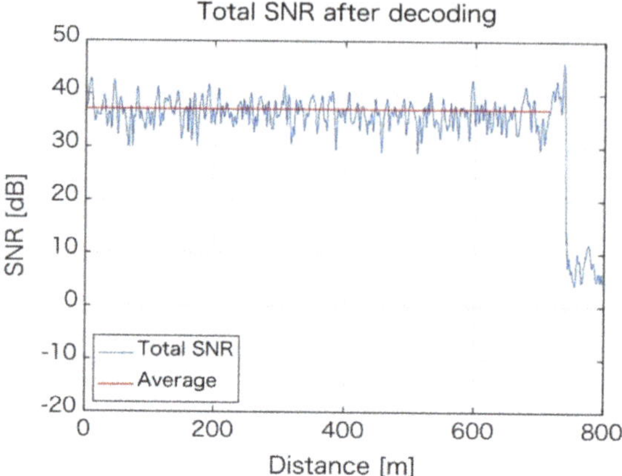

Figure 6. SNR distribution for the final (combined) data from all bands and polarization components.

In Figure 7, as the very small strain value is detectable, nano-strain, nε, are used. Please note that, during the tests, we were not always able to create a fully soundproof nor vibration-free environment and, thus, some signals at the low frequency region below 100 Hz, and at some other specific frequencies, were detected. Those noise peaks were not present in data sets acquired in oil and gas wells, thus they can be ignored when discussing the noise floor and performance.

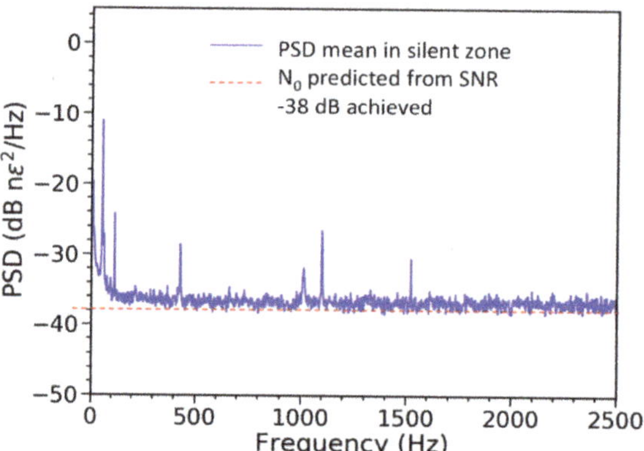

Figure 7. Noise power spectral density (PSD).

The noise floor of the instrument is at the level of $N_0 = -38$ nε^2 dB, with a 5 kS/s interrogation rate. This corresponds to $\sqrt{N_0} = 0.0126$ n$\varepsilon/\sqrt{\text{Hz}}$. This means that a sustained input of sound waves of given frequency with an amplitude of 12.6 pε, or larger, can be detected. Figure 8 provides a comparison of the detectable signal, noise floor, and strain amplitude in reference to other interrogators, including the system, which uses engineered fiber to increase the SNR. In this figure, GL stands for gauge length.

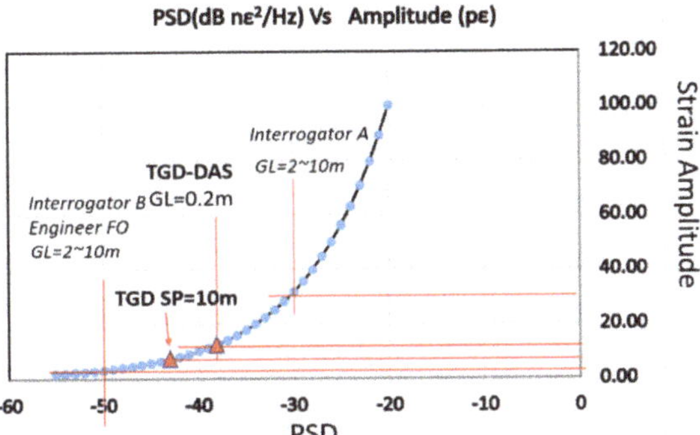

Figure 8. Performance of TGD-DAS in comparison with other reported results.

We should also note that, unlike time-domain DAS methods where gauge length and spatial resolution are generally equal, in the OFDR type DAS, the gauge length and spatial resolution have much wider ranges of valid values and can be freely selected.

In the results presented in this paper, a 500 MS/s digitizer was used. This sampling rate provides the best (smallest) gauge length of 20 cm. Such a gauge length can extend the measurable, maximum strain change in a single time step to 5.4 με; that is, two orders of magnitude higher than other available interrogators. This means the TGD-DAS can handle both extremes of the acoustic sensing in real-world applications, as it is capable of recording a strong vibration signal, and, at the same time, has the lowest noise floor in a standard SM fiber, and is able to record very weak signals.

Table 2 lists a few representative values for each variable. The actual performance is also linked to the interrogation rate, gauge length, and spatial resolution. Noting that the noise floor changes with these three quantities, this indicates that the design worked correctly. To the best of our knowledge, there is no other report that demonstrates this type of performance. The reasons for that might be signal-processing related, as usually papers and reports on DAS technology do not show the 'raw' interrogator performance, and include specialized processing steps before results are presented and passed to the user. Those additional processing steps might sometimes change the "engineering expectation". In this paper, all of the results are obtained by using FPGA and show reliable performance at the industrial level.

Table 2. Noise floor as a function of gauge length.

Interrogator Rate kS/s	Gauge Length m	Spatial Resolution m	Noise Floor		Maximum Measurable Strain με
			PSD dB nε²/Hz	Strain Amplitude pε	
2	0.2	1.8	−35	17.8	5.4
5	0.2	1.8	−38	12.6	5.4
5	3.0	3.0	−42	7.9	0.36
2	0.2	10.0	−45	5.6	5.4

One method to lower the noise floor in DAS measurements is by attempting to make Fiber Bragg Gratings (FBG) over the entire length of the optical fiber. This type of fiber is usually referred to as "engineered fiber". The minimum noise floor that could be achieved with that special optical fiber was 3 pε [8]. However, as listed in Table 2, noise floor of -42 nε^2 dB, corresponding to a detectable level of just 7.9 pε, can be obtained in standard SM with a gauge length of 3 m. This level is the highest ever achieved in the industry for standard SM fibers.

4. TGD-DAS Features, Advantages, and Future Improvements

In this section, we summarize the main advantages of TGD-DAS over other acoustic sensing methods and list improvements deployed in the NBX-S4000 interrogator.

4.1. Gauge Length and Spatial Resolution

Gauge length is defined as the interval length in the manufacturer's method of analyzing the phase [6]. The spatial sampling interval of NBX-S4000 is set to 20 cm, due to requirement of the frequency band to restore the chirp signal. For this purpose, the gauge length can be arbitrarily set at intervals, with the step of 20 cm, to analyze the acoustic signal.

The spatial resolution of the TGD-DAS is approximately 1.8 m, which is inversely proportional to the gauge length as shown in Table 2, and is determined by the pulse width of the chirped pulses after compression. A minimum gauge length of 20 cm results in a maximum permissible strain as high as 5.4 $\mu\epsilon$, which is 14 times larger than the maximum permissible strain at a gauge length of 1.8 m, which is a gauge length equal to the spatial resolution, as used in typical implementations. This ability to handle large strains is significant in practical terms. Moreover, even if the strain exceeds the maximum permissible strain, the "wrapped" phase difference can still be measured, tracked, and unwrapped (in the time direction), allowing one to detect sound waves with even larger amplitudes. The details of the phase unwrapping procedure must meet specific mathematical conditions, and those depend on the sampling frequency and the frequency of the sound waves. As this topic is beyond the scope of this paper, the details are omitted here.

4.2. Resolution Change by Post-Processing

The distinct feature of the TGD-DAS method is its ability to change the spatial resolution of the already acquired data by simple post-processing. In the current implementation, resolution can be changed from 1.8 to 13.5 m, simply by selecting the appropriate number of bands, instead of using the default three. We demonstrate that feature in the following application examples, where the same measurement data can be converted to phase results with different spatial resolutions.

4.3. Future Improvements

After carefully analyzing measurement errors in the prototype interrogator, we concluded that the main source of those errors is the frequency drift of the light source. The drift exhibits in some abnormal signal components, as shown in Figure 7. One of the methods to remove this error, we believe, could be to establish a pre-processing method for first estimating and then removing the frequency shift of the light source using an acoustic reference fiber.

5. User Interface and Output Design of NBX-S4000

The signal acquired by the interrogator unit can be automatically processed and used in events detection and a classification system. This requires a separate module for machine learning algorithm training. Figure 9 shows the overall scheme of automatic event detection as currently implemented in NBX-S4000.

An automatic monitoring system using acoustic signals in a wide range of applications and monitoring targets is an aspect of DAS that holds great promise. There are cases in which the AI module is executed in the acquisition mode (during signal acquisition) and some in the post-processing mode.

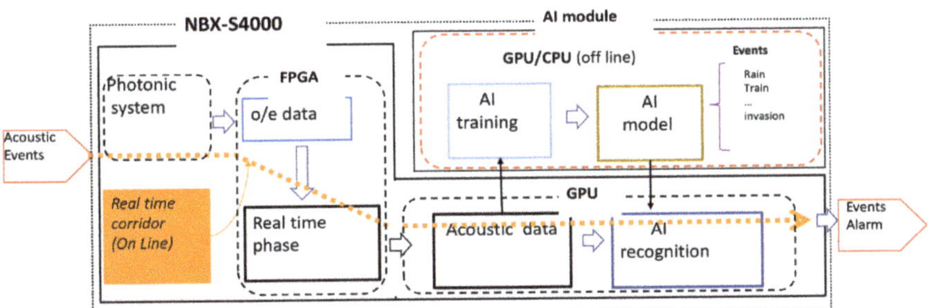

Figure 9. Events detection module, not an optic, but needed for AI processing.

6. Application Examples

This section presents a series of application examples, in which a developed TGD-DAS acquisition method and interrogator unit not only demonstrated their capabilities and potential, but also were also thoroughly tested in the field environment. Results from seismic applications, measurements of submarine cables, hydraulic fracturing in oil and gas, as well as intrusion detection, are discussed. We conclude this section by presenting one of the distinct features of TGD-DAS, namely, a posteriori resolution changes for already acquired data.

6.1. *Vertical Seismic Profiling (VSP) and Earthquakes Detection*

First, we present two applications related to detecting seismic events.

6.1.1. VSP

VSP is one of the most important techniques in geophysical exploration. It measures, using a set of sources and receivers, seismic wavefields to obtain properties of rock (such as velocity, impedance, attenuation, anisotropy). The seismic events are generated by the source, which is commonly a vibrator in onshore, and an air gun in offshore or marine environments. The receivers are usually placed along the depth (vertical axis) of the well.

There are many versions (configurations) of VSP type acquisitions, such as the zero-offset VSP, offset VSP, walkaway VSP, walk-above VSP, to name the most important ones. In the zero-offset VSP, the source is near the wellhead and above the vertical receiver array or sensing fiber. In the offset VSP, the source is located at some distance away from the vertical borehole. In the walk-away VSP, the source is moved in steps away from the well, while the receivers are held fixed. This approach effectively provides higher resolution seismic data and provides more continuous coverage than the standard offset VSP.

Until recently, geophones or accelerometers were used to record reflected seismic energy. With the increased sensitivity and accuracy of DAS, it became possible to obtain the VSP data using distributed sensing. This approach is preferable as it is economical (data are obtained faster and along the entire well) without need of moving the geophones array up and down the wellbore, and repeating the signal generation.

Figure 10a presents the results of measured data during walk-away VSP acquisition. The source was a small 15,000-pound P-wave vibrator, sweeping for 16 s over 10–150 Hz, linear. Four individual records were vertically stacked before cross-correlation and then cross-correlated with the pilot sweep to generate this record. The acquired data demonstrated the sensitivity and accuracy of TGD-DAS and the interrogator. The down-going and up-going reflected waves are clearly detected, providing data for further geological analysis. The obtained results were compared with geophones, confirming that the same level of accuracy was obtained, which means TGD-DAS can be used instead of geophones for VSP applications.

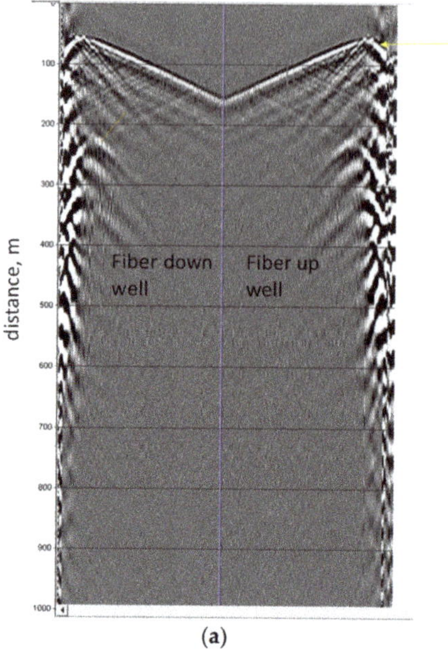

(a)

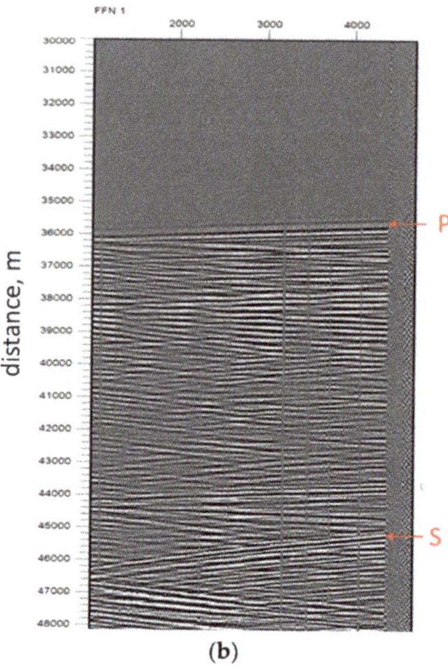

(b)

Figure 10. Seismic application examples. (**a**) Dataset acquired during walk-away VSP measurements in the vertical well; (**b**) seismic signal measured during the earthquake.

6.1.2. Earthquake Detection

The second example of seismic data, acquired using the NBX-4000 instrument, deals with earthquake measurements (as presented in Figure 10b). This dataset shows the earthquake-generated waves. The signal was recorded during submarine cable measurements, off the Japan coast. The results clearly demonstrate the sensitivity of the instrument, as both P- and S-waves are recorded. Those waves are annotated in Figure 10b. Moreover, the line-separating signal before and after the earthquake can be clearly observed, which proves the high reliability of the acquired data and results.

6.2. Long Distance Range–Submarine Cables Measurements

Acoustic sensing in optical fibers is emerging as one of the most important technologies in submarine seismic monitoring applications [9–11]. The sole reason is the fact that existing submarine infrastructure can be converted into seismic and wave sensors by simply connecting the interrogator unit to many (already deployed) optical fibers in seabed cables. Until very recently, the main challenge and problem faced when trying to deploy was the limited distance, which DAS interrogator units could cover. The submarine infrastructure and cables are very long and can be accessed (easily) only from the coast. To verify that the NBX-S4000 interrogator is capable of measuring submarine cables without the need for any amplifiers, the test in the long-term deep sea floor observatory off Cape Muroto, Shikoku Island, Japan, was performed. The distance range to cover (measure) was 80 km.

The following acquisition settings were used:

- Temporal sampling rate was set to 0.5 kS/s;
- Spatial resolution set to (standard) 1.8 m;
- Gauge length was 0.2 m.

The sampling rate was selected to ensure that there is only one pulse travelling at any given time in the 80 km long fiber. The data were streamed to permanent storage on disks, selecting proper regions of interest. An example of acquired data with recorded ocean wave propagation is shown in Figure 11, in which a 1 km long section, between 75 and 76 km of the cable, was extracted and plotted. The signal-to-noise ratio of the acquired signal was still high and indicated that NBX-S4000 can be deployed for such marine applications. The full results of those tests, including detected seismic events at seabed, will be reported and published separately.

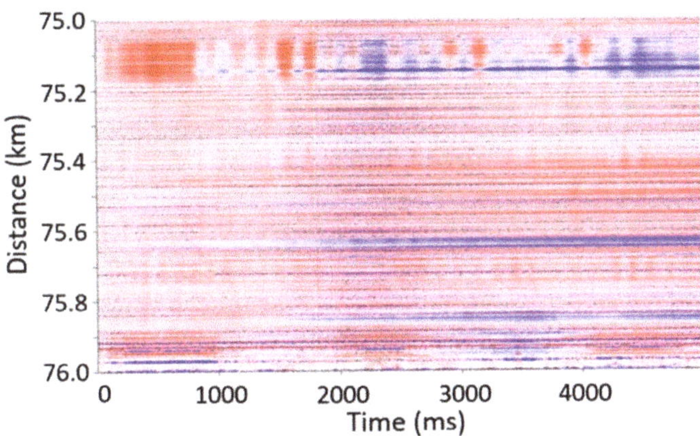

Figure 11. Observed acoustic signal in long-term deep sea floor observatory.

6.3. Hydraulic Fracturing and Fracture Hits Detection

Hydraulic fracturing, referred to as fracking, is a well stimulation process used in the oil and gas industry. It usually involves injecting water, sand, and chemicals under high pressure into a reservoir formation, of, generally, low-permeability rocks (e.g., shale). As a result, the fractures are created in the geological layer, and oil and/or gas can be extracted. The distributed sensing is particularly suited for monitoring of the fracking process, as it can provide real-time data, allowing engineering teams to ensure the correctness and state of the stimulation process, reached depths, and well integrity [8].

While measuring the signal on a fiber installed in well stimulation cannot be considered a great achievement, as the signal there is very strong, the detection of the signal on the neighboring well (referred to as a crosswell) certainly can. The ability to capture the signal there is of key interest for reservoir engineers, as it allows one to create the map of fracture networks and their azimuths, as well as to determine fracture geometry, spacing between them and far-field connectivity.

The NBX-S400 instrument was already used in both types of monitoring, in the stimulated well and in the crosswell. We present, in Figure 12, the normalized power of the "frac hits" detected on the crosswell. It clearly demonstrates the quality of acquired data, which were additionally validated by comparing them with results obtained using the Rayleigh Frequency Shift (RFS) method [12]. This is another acquisition technique, developed by Neubrex, which finds its use in various monitoring applications, including the oil and gas industry, due to its capability to detect strain changes, high-resolution, and sensitivity.

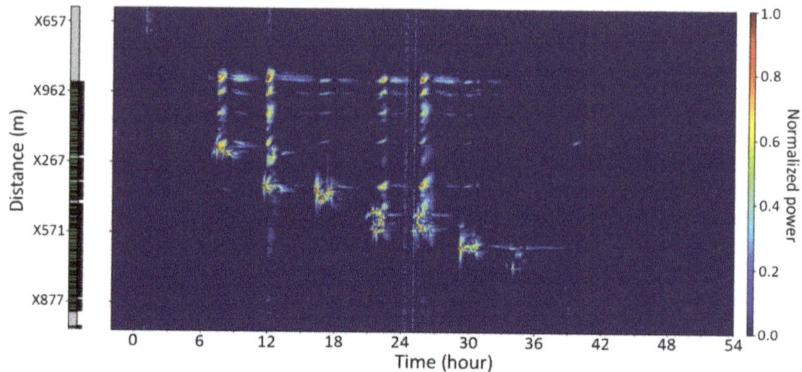

Figure 12. Signal detection (frac hits) on a crosswell.

6.4. Detection of Events Using "Dark Fibers"

The already built, extremely vast, telecommunication network of sub-surface fiber optics cables is commonly referred to as the "dark fiber" network [9]. While it serves its primary purpose by data transfer and communication, it can relatively (easily) convert into a sensing network, just by connecting it to the distributed sensing interrogator, which uses the available wavelength window and bandwidth. Such a sensing network can be used for earthquake detection and monitoring, structural health monitoring, or intrusion detection. In the latter, the system monitors the network for any malicious or unwanted events by performing continuous measurements. Generally, the detection system must also classify the signal; to do so, the signal quality must be high enough, not only to be detectable, but also to be distinguishable from other sources.

For intrusion detection on the railway tracks, a standard DAS instrument (which uses a single pulse technique) was previously used.

Figure 13a presents the location of the sensing fiber and the distances to the key points along the tracks. The attenuations of the acoustic signal, generated by a person walking along the tracks, at indicated key points P1, P2, and P3 are plotted in Figure 13a. While the standard instrument used during the test could only detect intrusion at locations P1 and, partially only, P2 and P3. SNR and noise floor level of the NBX-S300 instrument, using a standard single-pulse technique, was far too high for reliable intrusion detection. In Figure 13b, the noise level presented in this paper, the TGD-DAS instrument was also indicated. With noise floor at −37 dB, the signal from even the furthest distance from the sensing fiber would be easily detectable. The tests using the NBX-S4000 interrogator will be repeated at the earliest opportunity, as it requires coordination with a railway company.

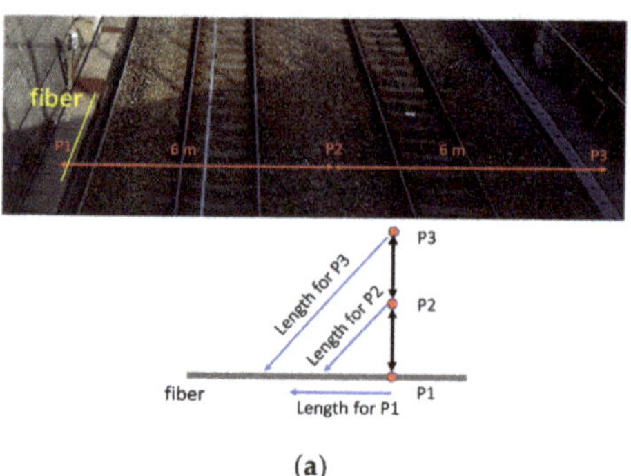

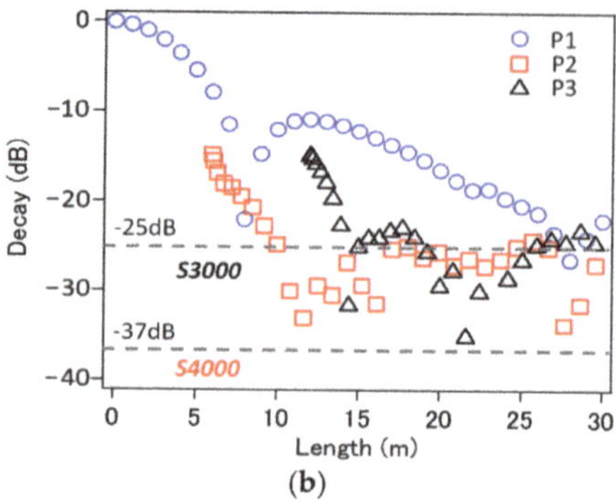

Figure 13. Intrusion detection example by NBX-S3000 (**a**) and (**b**) predicted improvement if NBX-S4000 is used.

6.5. Extraction of Dispersion Curve from DAS Seismic Wave Measurements Data

In critical infrastructure, such as dams, knowing and monitoring the geotechnical properties of subsoil layers plays an essential role in ensuring its safety. Several different methods can be used to estimate such properties. While intrusive methods, such as drilling, can sometimes be used, the surface methods are preferable, as they are low-cost, noninvasive, and environmentally friendly. What is also important, in many cases, including dams, the surface methods do not destroy or leave any lasting marks on the surface of the test site.

In surface wave methods, waves are generated and used to measure propagation velocity profiles as a function of depth. In dam embankment, this velocity is related to the moisture content in it and can be used to monitor any potential subsurface leaks. As a signal source, vehicle driving on top of the bank is used [13].

There are several methods proposed to extract the dispersion curve. Among them, the phase-shift method is regarded as effective and robust, and is widely used. This method can be applied, not only toward analysis of the surface waves, but also body waves.

The measured seismic profiles, including captured P- and S-waves, of the surface waves, are presented in Figure 14. The signal was obtained by stacking 35 datasets. The signal within the selected area (indicated by red rectangle) was then used as input data to the phase–shift dispersion curve analysis. The high quality of TGD-DAS (NBX-S4000) signal was compared directly with data obtained by geophones [14].

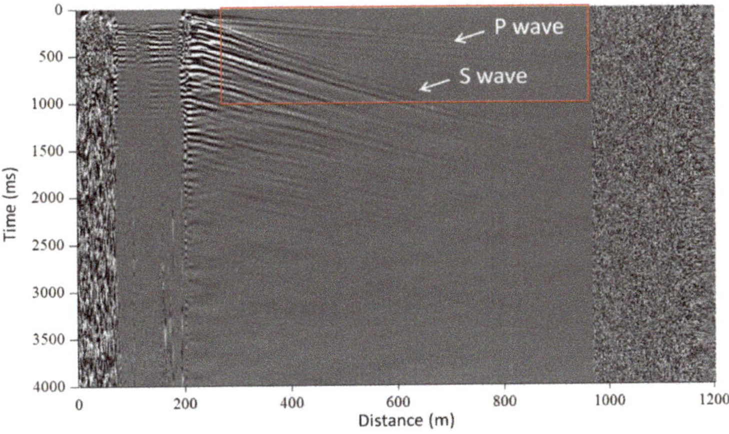

Figure 14. The measured seismic profiles of the surface wave.

First, during signal processing, the velocity-frequency map was generated (as shown in Figure 15a). The P-wave and S-wave are clearly seen in the figure. In the next step, peak values for each frequency line profile were determined, and the phase velocity of the P-wave and S-wave, as a function of frequency, are shown in Figure 15b.

We should note that, considering the very long lifetime of the optical cables and the high quality of the obtained signal and, therefore, the estimation results, it is possible to construct a permanent monitoring system using only sensing fiber and surface sources.

6.6. Variable Spatial Resolution for Already Acquired Data

Finally, we present one of the most distinct features of TGD-DAS, namely, its ability to modify the spatial resolution of the already acquired signal, simply by post-processing it. As described earlier in the paper, the spatial resolution in TGD-DAS depends on the selection of the frequency bands during the signal-decoding step. Hence, within limits determined by duration of the chirp, the spatial resolution can be changed. Currently supported are resolutions varying from 1.8 to 13.5 m.

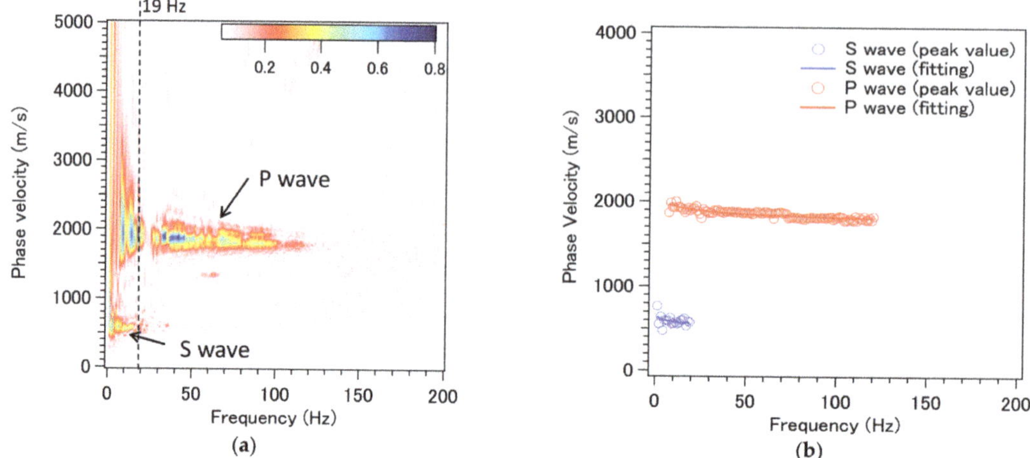

Figure 15. Velocity–frequency map (**a**) and dispersion curve (**b**).

The need to change the spatial resolution in some applications arises from the fact that characteristics of wave propagation in different mediums are different. The most common use is seismology. Here we present an example of such a resolution change for the same VSP dataset. Figure 16 shows the results with applied resolutions of 4, 7, and 10 m.

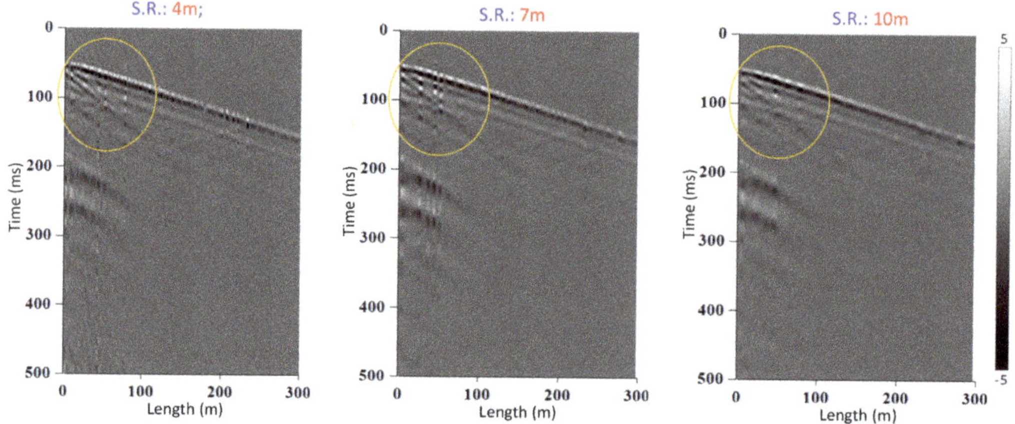

Figure 16. Results of signal processing with spatial resolution of 4, 7, and 10 m.

The spectral analysis of the obtained signal, presented in Figure 17a,b, confirms that lower (longer) resolutions had also lower noise floor. In this case, the signal peak value is the same, as results for different resolutions are obtained from the same raw data, and the noise floor can be reduced by almost 10 dB.

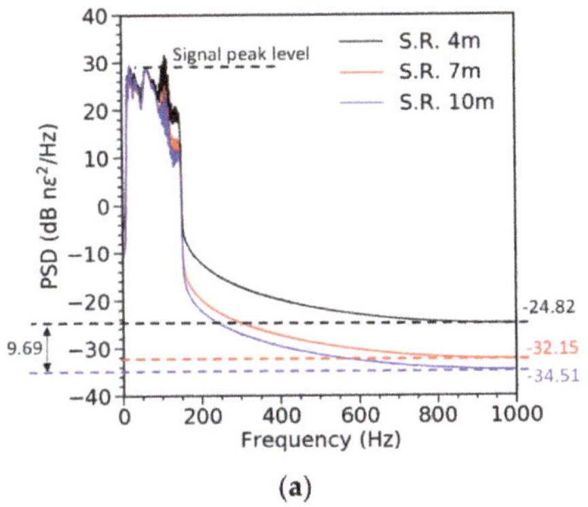

(a)

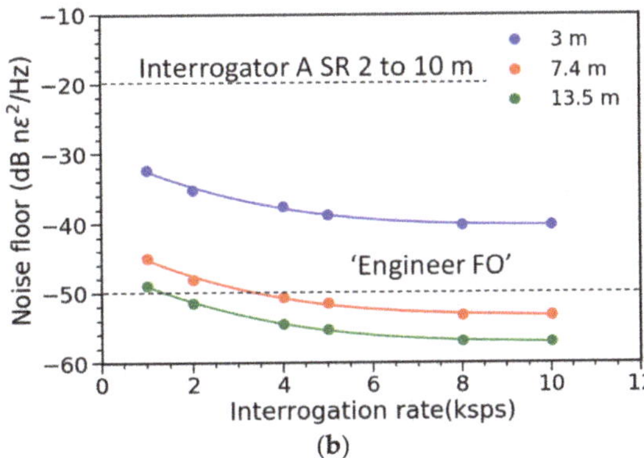

(b)

Figure 17. Spectral analysis results (**a**) with spatial resolution (SR) of 4, 7, and 10 m at interrogator rate of 2 kS/s (**b**) Performance for different interrogation rate.

Further analysis of the noise floor as a function of the interrogation rate and applied spatial resolution clearly demonstrates that performance of TGD-OFDR in standard SM fibers not only matches, but also surpasses that obtained by systems, depending on the specialty (weak FBG), engineered fibers. The deployment of engineered fibers has very practical implications, as it prevents from using the fiber for other sensing purposes and interrogators, so standard fibers are preferable.

Modifiable spatial resolution feature of TGD-DAS is particularly useful in any seismic application, as acquired data can be used to extract different features of the signal a posteriori, instead of determining them before any signal is measured.

7. Discussion

The DAS method and its implementation presented in this paper, compared to previous studies and applications, exhibit several substantial improvements and advantages. The most notable is that the synthesis of signals from extracted frequency bands, for both polarization components we detect, eliminates any phase fading from the acquired signal. Hence, the inherent problem in all OTDR-based systems, resulting in singularity when determining the phase value, is eliminated, simplifying the signal-processing part. Another substantial improvement is that spatial resolution is dependent on frequency range rather than pulse duration. This means that resolution can be modified after the signal is already acquired, and the resolution is independent of the measured fiber length. This feature makes TGD-DAS substantially different from any OTDR-based method. Results presented in the paper used varying resolutions between 1.8 and 13.5 m. We also presented in the field measurements, without any additional amplification, covering a distance of more than 80 km. In summary, the method, and its deployment in the NBX-S4000 interrogator, allows one to generate the signal with an optical power of a 200 m long pulse, but kept in a 2-m spatial resolution range.

8. Conclusions

This paper presents a practical DAS system that utilizes the advantages of both the frequency and time domain signal acquisition and processing. The TGD-DAS method, as well as the interrogator unit, achieves high precision, high spatial-resolution acoustic signal detection by frequency-division multiplexing, and real-time signal processing. The developed technique has room for further improvement, by expanding it to a wider frequency range and longer measurement distances. Those issues will be addressed in the future. The results obtained during the research presented in this paper revealed that performance and accuracy improvements are possible by developing a better light source and suppressing the noise caused by its varying frequency.

Author Contributions: Conceptualization, K.K. and Q.L.; methodology, K.K.; software, A.G.; validation, D.A., C.-H.L. and K.N.; formal analysis, K.N.; investigation, D.A.; resources, K.K.; data curation, A.G.; writing—original draft preparation, A.G.; writing—review and editing, K.K.; visualization, A.G. and D.A.; supervision, K.K. and Z.H. All authors have read and agreed to the published version of the manuscript.

Funding: This research received no external funding.

Institutional Review Board Statement: Not applicable.

Informed Consent Statement: Not applicable.

Acknowledgments: The authors thank the Neubrex technical team for their help during this research.

Conflicts of Interest: The authors declare no conflict of interest.

References

1. Taylor, H.F.; Lee, C.E. Apparatus and Method for Fiber Optic Intrusion Sensing. U.S. Patent 5194847, 16 March 1993.
2. Kishida, K.; Guzik, A.; Li, T.; Nishiguchi, K.; Azuma, D.; Liu, Q.; He, Z. Commercialization of real-time distributed acoustic fiber optic sensing (DAS) utilizing chirped pulse. *IEICE Tech. Rep.* **2020**, *120*, 39–44.
3. Nishiguchi, K.; Li, T.; Guzik, A.; Yokoyama, M.; Kishida, K. Fabrication of a Distributed Acoustic Sensor Using Optical Fiber and Its Signal Processing. *IEICE Tech. Rep.* **2015**, *115*, 29–34.
4. Liu, Q.; Fan, X.; He, Z. Time-gated digital optical frequency domain reflectometry with 1.6-m spatial resolution over entire 110-km range. *Opt. Express* **2015**, *23*, 25988–25995. [CrossRef] [PubMed]
5. Liu, Q.; Liu, L.; Fan, X.; Du, J.; Ma, L.; He, Z. A Novel Optical Fiber Reflectometry Technique with High Spatial Resolution and Long Distance. In Proceedings of the Asian Communications and Photonics Conference, Shanghai, China, 11–14 November 2014.
6. SEAFOM. *Document—02 (SEAFOM MSP-02). Measuring Sensor Performance, DAS Parameter Definitions and Tests*; SEAFOM: Epsom, UK, 2018.
7. Chen, D.; Liu, Q.; He, Z. Phase-detection distributed fiber-optic vibration sensor without fading-noise based on time-gated digital OFDR. *Opt. Express* **2017**, *25*, 8315–8325. [CrossRef] [PubMed]
8. Richter, P.; Parker, T.; Woerpel, C.; Wu, Y.; Rufino, R.; Farhadiroushan, M. Hydraulic fracture monitoring and optimization in unconventional completions using a high-resolution engineered fibre-optic Distributed Acoustic Sensor. *First Break* **2019**, *37*, 63–68. [CrossRef]
9. Kishida, K.; Araki, E.; Azuma, D.; Yokobiki, T.; Matsumoto, H.; Guzik, A. Measurements of 80 km long submarine cables using TGD-DAS system. In Proceedings of the EAGE Workshop on Fiber Optic Sensing for Energy Applications in Asia Pacific, Kuala Lumpur, Malaysia, 9–11 November 2020; European Association of Geoscientists and Engineers: Houten, The Netherlands, 2020; Volume 2020, pp. 1–5.
10. Ajo-Franklin, J.; Dou, S.; Lindsey, N.J.; Monga, I.; Tracy, C.; Robertson, M.; Tribaldos, V.R.; Ulrich, C.; Freifeld, B.; Daley, T.; et al. Distributed Acoustic Sensing Using Dark Fiber for Near-Surface Characterization and Broadband Seismic Event Detection. *Sci. Rep.* **2019**, *9*, 1–14. [CrossRef] [PubMed]
11. Matsumoto, H.; Araki, E.; Kimura, T.; Fujie, G.; Shirasishi, K.; Tonegawa, T.; Obana, K.; Arai, R.; Kaiho, Y.; Nakamura, Y.; et al. Detection of hydroacoustic signals on a fiber-optic submarine cable. *Sci. Rep.* **2021**, *11*, 1–12. [CrossRef]
12. Kishida, K.; Yamauchi, Y.; Guzik, A. Study of Optical Fibers Strain-Temperature Sensitivities Using Hybrid Brillouin-Rayleigh System. *Photonic Sens.* **2014**, *4*, 1–11. [CrossRef]
13. Kurihara, K. Application of Distributed Acoustic Sensing (DAS) with Optical Fibers to Seismic Surveys. In Proceedings of the 75th Annual Conference of JSCE, Virtual Conference, 2 November 2020; p. 428.
14. Onishi, K.; Inazaki, F.; Ogawara, T.; Kobayashi, T.; Nishizawa, O. Analysis of Surface Waves and Propagation Speeds Recorded by Distributed Acoustic Sensing. In Proceedings of the 139th conference of the Society of Exploration Geophysicists of Japan, Toyama International Conference Center, Toyama, Japan, 22–24 October 2018; pp. 107–110.

MDPI
St. Alban-Anlage 66
4052 Basel
Switzerland
Tel. +41 61 683 77 34
Fax +41 61 302 89 18
www.mdpi.com

Sensors Editorial Office
E-mail: sensors@mdpi.com
www.mdpi.com/journal/sensors

www.ingramcontent.com/pod-product-compliance
Lightning Source LLC
LaVergne TN
LVHW070634170726
843515LV00004B/246

* 9 7 8 3 0 3 6 5 4 1 5 7 0 *